COMPUTER METHODS FOR MATHEMATICAL COMPUTATIONS

Prentice-Hall
Series in Automatic Computation

MARTIN, *Principles of Data-Base Management*
MARTIN, *Programming Real-Time Computing Systems*
MARTIN, *Security, Accuracy, and Privacy in Computer Systems*
MARTIN, *Systems Analysis for Data Transmission*
MARTIN, *Telecommunications and the Computer*
MARTIN, *Teleprocessing Network Organization*
MARTIN AND NORMAN, *The Computerized Society*
MCKEEMAN, et al., *A Compiler Generator*
MEYERS, *Time-Sharing Computation in the Social Sciences*
MINSKY, *Computation: Finite and Infinite Machines*
NIEVERGELT, et al., *Computer Approaches to Mathematical Problems*
PLANE AND MCMILLAN, *Discrete Optimization:*
    Integer Programming and Network Analysis for Management Decisions
POLIVKA AND PAKIN, *APL: The Language and Its Usage*
PRITSKER AND KIVIAT, *Simulation with GASP II:*
    A FORTRAN-based Simulation Language
PYLYSHYN, ed., *Perspectives on the Computer Revolution*
RICH, *Internal Sorting Methods Illustrated with PL/1 Programs*
RUDD, *Assembly Language Programming and the IBM 360 and 370 Computers*
SACKMAN AND CITRENBAUM, eds., *On-Line Planning:*
    Towards Creative Problem-Solving
SALTON, ed., *The SMART Retrieval System:*
    Experiments in Automatic Document Processing
SAMMET, *Programming Languages: History and Fundamentals*
SCHAEFER, *A Mathematical Theory of Global Program Optimization*
SCHULTZ, *Spline Analysis*
SCHWARZ, et al., *Numerical Analysis of Symmetric Matrices*
SHAH, *Engineering Simulation Using Small Scientific Computers*
SHAW, *The Logical Design of Operating Systems*
SHERMAN, *Techniques in Computer Programming*
SIMON AND SIKLOSSY, eds., *Representation and Meaning:*
    Experiments with Information Processing Systems
STERBENZ, *Floating-Point Computation*
STOUTEMYER, *PL/I Programming for Engineering and Science*
STRANG AND FIX, *An Analysis of the Finite Element Method*
STROUD, *Approximate Calculation of Multiple Integrals*
TANENBAUM, *Structured Computer Organization*
TAVISS, ed., *The Computer Impact*
UHR, *Pattern Recognition, Learning, and Thought:*
    Computer-Programmed Models of Higher Mental Processes
VAN TASSEL, *Computer Security Management*
VARGA, *Matrix Iterative Analysis*
WAITE, *Implementing Software for Non-Numeric Application*
WILKINSON, *Rounding Errors in Algebraic Processes*
WIRTH, *Algorithms + Data Structures = Programs*
WIRTH, *Systematic Programming: An Introduction*
YEH, ed., *Applied Computation Theory: Analysis, Design, Modeling*

COMPUTER METHODS FOR
MATHEMATICAL COMPUTATIONS

GEORGE E. FORSYTHE

MICHAEL A. MALCOLM

Department of Computer Science
University of Waterloo

CLEVE B. MOLER

Department of Mathematics and Statistics
University of New Mexico

PRENTICE-HALL, INC.

ENGLEWOOD CLIFFS, N. J. 07632

Library of Congress Cataloging in Publication Data

FORSYTHE, GEORGE, ELMER, (date)
    Computer methods for mathematical computations.

    (Prentice-Hall series in automatic computation)
    Bibliography: p.
    Includes index.
    1. Numerical analysis—Data processing.
2. FORTRAN (Computer program language) I. Malcolm,
Michael A., (date) joint author. II. Moler,
Cleve B., joint author. III. Title.
QA297.F568   519.4   76–30819
ISBN 0–13–165332–6

© 1977 by Prentice-Hall, Inc., Englewood Cliffs, New Jersey 07632

10

Printed in the United States of America

PRENTICE-HALL INTERNATIONAL, INC., *London*
PRENTICE-HALL OF AUSTRALIA PTY. LIMITED, *Sydney*
PRENTICE-HALL OF CANADA, LTD., *Toronto*
PRENTICE-HALL OF INDIA PRIVATE LIMITED, *New Delhi*
PRENTICE-HALL OF JAPAN, INC., *Tokyo*
PRENTICE-HALL OF SOUTHEAST ASIA PTE. LTD., *Singapore*
WHITEHALL BOOKS LIMITED, *Wellington, New Zealand*

CONTENTS

*There are, so to speak, in the
mathematical country, precipices
and pit-shafts down which it would
be possible to fall, but that need not
deter us from walking about.*

L. F. Richardson

PREFACE

This book evolved from a set of notes used in introductory numerical methods courses at Stanford University, the University of New Mexico and the University of Waterloo. These universities have two such courses, titled something like "Numerical Computing" and "Numerical Analysis". The numerical computing course is for students of science and engineering who simply want to use numerical methods in their work. It lasts only one term and has minimal mathematical prerequisites. The numerical analysis course covers essentially the same methods, but lasts a full year. The additional time is devoted to deeper mathematical analysis. This book is intended primarily for the numerical computing course, but could also be used to supplement a more theoretical text in the numerical analysis course.

Ten Fortran subroutines are included in the book. It is expected that they will be used extensively both during and after the course. The programs have been carefully written so that they may be easily used on a wide variety of computers with little or no modification.

One of our colleagues has pointed out that "it is an order of magnitude easier to write two good subroutines than it is to decide which one is best". In choosing among the various subroutines available for a particular problem, we have placed considerable emphasis on the clarity and style of programming. If several subroutines have comparable accuracy, reliability and efficiency, we have chosen the one which is the least difficult to read and use.

In recent years, there has been increased recognition of the importance of writing, testing and distributing quality mathematical software. So, although we believe that our subroutines are among the best available at the time the book is being written, we hope and expect that they will eventually be superceded. One of our goals in writing the book is to help the reader look for better subroutines and recognize and appreciate them when they become available.

A copy of the subroutines in 1966 ANSI Standard Fortran is available from the authors. Both single and double precision versions are provided on a single tape. The tape is provided in two formats: 7-track, BCD, 556 BPI, and 9-track, EBCDIC, 800 BPI. A deck containing about 2000 cards is also available upon special request. Write to: Prof. Cleve Moler, Department of Mathematics, University of New Mexico, Albuquerque, N.M. 87131.

We are indebted to many people for help and encouragement during the preparation of the manuscript. Following is a list of some of those people: Richard Allen, Mary Bodley, John Bolstad, Erin Brent, Richard Brent, Francena Brumbaugh, Andy Conn, Carl deBoor, Fred Dorr, C. William Gear, Keith Geddes, Alan George, Gene Golub, Dennis Jerpersen, Rondall Jones, William Kahan, David Kahaner, Shaaron Kent, Fred Krogh, Doug Lawson, James Lyness, Kathryn Moler, Teresa Moler, Walter Murray, Larry Nazareth, Joe Oliger, Victor Pereyra, Steve Pruess, George Ramos, Laurie Rogers, Michael Saunders, Larry Shampine, Laura Staggers, G. W. Stewart, Mike Steuerwalt, David Stoutemyer, Richard Underwood, H. A. Watts, and Michael Wester. Thanks to Masatake Mori and Robert Steel who pointed out errors in the first printing. Special thanks are due to the many students who suffered through early versions of the notes and helped debug the sub-routines as well.

Prior to the completion of this manuscript, on April 9, 1972, Professor George E. Forsythe met an untimely death at the age of 55. At the time, Professor Forsythe was the chairman and founder of Stanford's Computer Science Department. His loss was mourned by the academic and scientific communities, while we, George's friends, colleagues, and students, felt a very personal loss. In his tribute given at the memorial service, Professor Edward A. Feigenbaum spoke for many of us when he said of Professor Forsythe: "He was one of the gentlest and most humane of people, and we loved him for it . . . He was a man without violence of the spirit . . ."

The spirit of this book is George Forsythe's. It is our desire that it find peaceful uses worthy of his memory.

MICHAEL A. MALCOLM
University of Waterloo

CLEVE B. MOLER
University of New Mexico

1 INTRODUCTION

This book is concerned with solving mathematical problems using automatic digital computers. The reader is assumed to have completed two years of university mathematics. This should include differential and integral calculus and a little matrix theory and differential equations. He is also assumed to have access to a medium- or large-scale scientific computer and be able to program it using the local dialect of Fortran.

A very important part of the book is a set of Fortran subroutines. In fact, the book might well be regarded as an extensive "user's guide" for the subroutines. These are *not* the simple, illustrative sample programs found in many textbooks but rather are representative of the state of the art in current scientific computing. We intend that the Fortran subroutines be read by human readers as well as machines; that is why they are printed in the text. We also intend that the programs be used to solve problems given in the text.

We have followed a rather unconventional philosophy with regard to the presentation of material. Most engineers and scientists have neither the time nor the inclination to keep abreast of current literature in numerical analysis. We have observed that many people solving practical numerical problems rely upon the methods, if not the programs, used in previous course work. Therefore, we decided to present up-to-date subroutines for solving standard problems of mathematics.

Many of these subroutines are very intricate, and to describe in detail how and why they work would require much longer than most students can spend on this type of material. Hence, it is necessary to treat many of these subroutines as *black boxes*. Treating subroutines as black boxes is a common practice. The hardware subroutines for such things as addition and multiplication are treated as black boxes by most computer programmers. Few people really understand how they work. For the past several years, subroutines used to compute standard functions of analysis (e.g., sin, cos, etc.) have been treated as black boxes, and few programmers understand them in any detail. Such subroutines are accepted as primitives which are expected to work. It is the opinion of the authors that it is now time for programmers to accept subroutines which solve linear systems, ordinary differential equations, and many other standard mathematical problems as primitives which (usually) work.

However, it is not sufficient to simply explain how to call each of the subroutines and say that they "usually work." There is a wealth of pitfalls in numerical computation. The student must be properly warned of these pitfalls. He should learn to look for symptoms of numerical ill health and to correctly diagnose the problem(s). This requires a certain level of understanding of the numerical methods employed by the various subroutines. The programmer will also find occasion to modify a subroutine or use a variation of a method to solve a related problem. This usually requires more detailed knowledge.

We have tried to compromise on the amount of detail included with each method. The student may come to regard the subroutines as "gray boxes" rather than black. Where appropriate, we have referenced the literature for further discussion of a particular topic.

A short discussion of various decisions and style considerations that went into the Fortran subroutines is included in Section 1.2 of this chapter.

Since we shall be able to cover only a small part of the subject, we shall rely heavily on references to other books and articles. So, we begin our presentation with the bibliography.

1.1. BIBLIOGRAPHY

Two introductory books, mainly intended for the desk-machine user but having a substantial carry-over to automatic computing:

HILDEBRAND, F. B. (1974), *Introduction to numerical analysis*, 2nd ed. New York: McGraw-Hill.

LANCZOS, C. (1957), *Applied analysis*. Englewood Cliffs, N.J.: Prentice-Hall.

Some general books, in which automatic computers are considered:

ACTON, F. S. (1970), *Numerical methods that (usually) work*. New York: Harper & Row. (Two especially good chapters on removing singularities.)

AMES, W. F. (1969), *Numerical methods for partial differential equations*. New York: Barnes & Noble.

BEREZIN, I. S., and N. P. ZHIDKOV (1965), *Computing methods*, 2 vols. Elmsford, N.Y.: Pergamon. (A translation by Blunn of the 1959 textbook at Moscow State University for a two-year course.)

CONTE, S. D., and C. DEBOOR (1972), *Elementary numerical analysis; an algorithmic approach*, 2nd ed. New York: McGraw-Hill.

DAHLQUIST, G., and Å. BJÖRCK (1974), *Numerical methods*, Englewood Cliffs, N.J.: Prentice-Hall. (Complete, up-to-date text in numerical analysis.)

FOX, L., and D. F. MAYERS (1968), *Computing methods for scientists and engineers*. Oxford: Clarendon Press.

HAMMING, R. W. (1971), *Introduction to applied numerical analysis*. New York: McGraw-Hill.

HAMMING, R. W. (1973), *Numerical methods for scientists and engineers*, 2nd ed. New York: McGraw-Hill. (Hamming wants you to combine mathematical theory, heuristical analysis, and computer methods in the solution of your problem: "The purpose of computing is insight, not numbers.")

HENRICI, P. (1964), *Elements of numerical analysis*. New York: Wiley. (Elementary fundamentals, done very carefully. Nothing on matrix problems.)

MCCRACKEN, D. D., and W. S. DORN (1964), *Numerical methods and Fortran programming*. New York: Wiley.

RALSTON, A. (1965), *A first course in numerical analysis*. New York: McGraw-Hill. (In contrast to Henrici and Hamming, Ralston has a substantial amount on computational methods of linear algebra.)

RALSTON, A., and H. S. WILF (1960, 1967), *Mathematical methods for digital computers*, vol. 1 and vol. 2. New York: Wiley. (Mathematical descriptions and flowcharts of algorithms of varying quality. Those in Volume 2 are generally much better.)

SHAMPINE, L., and R. ALLEN (1973), *Numerical computing*. Philadelphia: Saunders. (A text similar in spirit to this one. Contains several good Fortran subroutines.)

TODD, J. (editor) (1962), *Survey of numerical analysis*. New York: McGraw-Hill.

WENDROFF, B. (1969), *First principles of numerical analysis. An undergraduate text*. Reading, Mass.: Addison-Wesley.

Bibliography of tables:

FLETCHER, A., J. C. P. MILLER, and L. ROSENHEAD (1962), *Index of mathematical tables*, 2nd ed., 2 vols. Oxford: Blackwell's. (This lists nearly every published table of mathematical functions up to around 1958 to 1959, and there were many. Indispensable.)

Lists of elementary functions that approximate transcendental functions:

HART, J. F., et al. (1968), *Computer approximations*. New York: Wiley.

HASTINGS, C., JR., et al. (1955), *Approximations for digital computers*. Princeton, N.J.: Princeton University Press. (Formulas with graphs of error curves. Most computer library routines are based on these formulas or improvements thereof.)

LYUSTERNIK, L. A., O. A. CHERVONENKIS, and A. R. YANPOLŚKII (1965), *Handbook for computing elementary functions*. Elmsford, N.Y.: Pergamon. (Translation by G. Tee of the 1963 Russian monograph.)

Best buy:

ABRAMOWITZ, M., and I. STEGUN (editors) (1964), *Handbook of mathematical functions with formulas, graphs, and mathematical tables*. National Bureau of Standards Applied Mathematics Series, vol. 55, Washington, D.C.: Government Printing Office. (Dover, New York, has issued a paperbound reprint. Everything for the occasional computer and user of special functions. Over 150,000 copies sold.)

Some advanced and specialized books on numerical methods:

AHLBERG, J. H., E. N. NILSON, and J. L. WALSH (1967), *The theory of splines and their applications*. New York: Academic Press.

BABUŠKA, I., M. PRAGER, and E. VITÁSEK (1966), *Numerical processes in differential equations*. New York: Wiley-Interscience. (Translation by M. Borůvková.)

BLACKMAN, R. B., and J. W. TUKEY (1958), *The measurement of power spectra*. New York: Dover.

BRENT, R. P. (1972), *Algorithms for minimization without derivatives*. Englewood Cliffs, N.J.: Prentice-Hall.

COLLATZ, L. (1966), *The numerical treatment of differential equations*, 3rd ed. Berlin: Springer. (Translation from German 2nd ed. by P. G. Williams.)

COLLATZ, L. (1966), *Functional analysis and numerical mathematics*. New York: Academic Press. (Applications of functional analysis to computational problems. Translation from German edition of 1964 by Hansjörg, Oser.)

DANTZIG, G. B. (1963), *Linear programming and extensions*. Princeton, N.J.: Princeton University Press.

DAVIS, P. J. (1963), *Interpolation and approximation*. Waltham, Mass.: Ginn/ Blaisdell.

DAVIS, P. J., and P. RABINOWITZ (1975), *Methods of numerical integration*. New York: Academic Press.

DEKKER, T. J. (1968), *Algol 60 procedures in numerical algebra, Part 1; Part 2* (with W. Hoffmann). Amsterdam: Mathematisch Centrum.

FADDEEV, D. K., and V. N. FADDEEVA (1963), *Computational methods of linear algebra*. New York: Dover.

FORSYTHE, G. E., and C. B. MOLER (1967), *Computer solution of linear algebraic systems*. Englewood Cliffs, N.J.: Prentice-Hall. (Programs in Algol, Fortran, and PL/1 for solving linear equation systems, with discussion of relevant theory.)

FORSYTHE, G. E., and W. R. WASSOW (1960), *Finite-difference methods for partial differential equations*. New York: Wiley.

FOX, L. (1957), *The numerical solution of two-point boundary problems.* Oxford: Clarendon Press.

GEAR, C. W. (1971), *Numerical initial value problems in ordinary differential equations.* Englewood Cliffs, N.J.: Prentice-Hall.

GILL, P., and W. MURRAY (1975), *Constrained optimization.* New York: Academic Press.

GREENSPAN, D. (1968), *Lectures on the numerical solution of linear, singular and nonlinear differential equations.* Englewood Cliffs, N.J.: Prentice-Hall.

GREGORY, R. T., and L. K. KARNEY (1969), *A collection of matrices for testing computational algorithms.* New York: Wiley-Interscience.

HAMMERSLEY, J. M., and D. C. HANDSCOMB (1964), *Monte Carlo methods.* London: Methuen. (Method of pseudorandom sampling to estimate numerical quantities.)

HANDSCOMB, D. C. (editor), (1966), *Methods of numerical approximation.* Elmsford, N.Y.: Pergamon. (Modern. Excellent bibliography.)

HENRICI, P. (1962), *Discrete-variable methods in ordinary differential equations.* New York: Wiley.

HOUSEHOLDER, A. S. (1964), *The theory of matrices in numerical analysis.* Waltham, Mass.: Ginn/Blaisdell.

HOUSEHOLDER, A. S. (1970), *The numerical treatment of a single nonlinear equation.* New York: McGraw-Hill.

ISAACSON, E., and H. B. KELLER (1966), *Analysis of numerical methods.* New York: Wiley. (Textbook on numerical analysis at beginning graduate level.)

KELLER, H. B. (1968), *Numerical methods for two-point boundary value problems.* Waltham, Mass.: Ginn/Blaisdell.

KNUTH, D. E. (1969), *The art of computer programming*, vol. 2: "Seminumerical algorithms." Reading, Mass.: Addison-Wesley. (See Chapter 3 on random numbers and Chapter 4 on floating-point arithmetic.)

KOWALIK, J., and M. R. OSBORNE (1968), *Methods for unconstrained optimization problems.* New York: American Elsevier.

KRYLOV, V. I. (1962), *Approximate calculation of integrals.* New York: Macmillan. (Translation by A. H. Stroud.)

LASDON, L. S. (1970), *Optimization theory for large systems.* New York: Macmillan.

LAWSON, C. L., and R. J. HANSON (1974), *Solving least squares problems.* Englewood Cliffs. N.J.: Prentice-Hall.

MOORE, R. E. (1966), *Interval analysis.* Englewood Cliffs, N.J.: Prentice-Hall. (Theory of replacing arithmetic of numbers by arithmetic of real intervals, a form of automated error analysis.)

MURRAY, W. (1972), *Numerical methods for unconstrained optimization.* New York: Academic Press.

ORTEGA, J. M., and W. C. RHEINBOLDT (1960), *Iterative solution of nonlinear equations in several variables*. New York: Academic Press.

ORTEGA, J. M. (1972), *Numerical analysis, a second course*. New York: Academic Press.

REID, J. K. (editor) (1971), *Large sparse sets of linear equations*. New York: Academic Press.

RHEINBOLDT, W. C. (1974), *Methods for solving systems of nonlinear equations*. Philadelphia: SIAM.

RICE, J. R. (1964), *The approximation of functions*, vol. 1: "Linear theory"; vol. 2: "Nonlinear and multivariate theory." Reading, Mass.: Addison-Wesley.

RICE, J. R. (1971), *Mathematical software*. New York: Academic Press.

RICHTMYER, R. D., and K. W. MORTON (1967), *Difference methods for initial-value problems*, 2nd ed. New York: Wiley-Interscience.

RIVLIN, T. J. (1969), *An introduction to the approximation of functions*. Waltham, Mass.: Ginn/Blaisdell.

ROSE, D. J., and R. A. WILLOUGHBY (editors) (1972), *Sparse matrices and their applications*. New York: Plenum.

SAUL'EV, V. K. (1964), *Integration of equations of parabolic type by the methods of nets*. Elmsford, N.Y.: Pergamon. (Translation by G. J. Tee.)

SCHULTZ, M. (1973), *Spline analysis*. Englewood Cliffs, N.J.: Prentice-Hall.

SHAMPINE, L. F., and M. K. GORDON (1975), *Computer solution of ordinary differential equations*. San Francisco: W. H. Freeman.

SMITH, B. T., J. M. BOYLE, B. S. GARBOW, Y. IKEBE, V. C. KLEMA, and C. B. MOLER (1974), *Matrix eigensystem routines—EISPACK guide*. Berlin: Springer.

SNYDER, M. A. (1966), *Chebyshev methods in numerical approximation*. Englewood Cliffs, N.J.: Prentice-Hall.

STEWART, G. W. (1973), *Introduction to matrix computations*. New York: Academic Press.

STRANG, W. G., and G. J. FIX (1973), *An analysis of the finite element method*. Englewood Cliffs, N.J.: Prentice-Hall.

STROUD, A. H. (1971), *Approximate calculation of multiple integrals*. Englewood Cliffs, N.J.: Prentice-Hall.

STROUD, A. H., and D. SECREST (1966), *Gaussian quadrature formulas*. Englewood Cliffs, N.J.: Prentice-Hall.

TRAUB, J. F. (1964), *Iterative methods for the solution of equations*. Englewood Cliffs, N.J.: Prentice-Hall.

VARGA, R. S. (1962), *Matrix iterative analysis*. Englewood Cliffs, N.J.: Prentice-Hall. (This is especially devoted to the solution of large partial differential equations of the types appearing in reactor calculations, with special discussion of implicit methods.)

WALSH, J. (editor) (1967), *Numerical analysis, an introduction.* Washington, D.C.: Thompson Book Co.

WILKINSON, J. H. (1963), *Rounding errors in algebraic processes.* Englewood Cliffs, N.J.: Prentice-Hall. (Inverse roundoff analysis.)

WILKINSON, J. H. (1965), *The algebraic eigenvalue problem.* New York: Oxford University Press. (The definitive treatise.)

WILKINSON, J. H., and C. REINSCH (1971), *Linear algebra, handbook for automatic computation,* vol. 2. Berlin: Springer.

YOUNG, D. M. (1971), *Iterative solution of large linear systems.* New York: Academic Press.

Selected journals:

ACM Transactions on Mathematical Software.

BIT. (Scandinavian, most articles in English.)

Communications of the ACM (Association for Computing Machinery).

The Computer Journal. (Published by the British Computer Society.)

International Journal for Numerical Methods in Engineering.

Journal of Computational Physics.

Journal of the Institute for Mathematics and Its Applications (British).

Mathematics of Computation (American Mathematical Society).

Numerische Mathematik. (German, most articles in English.)

SIAM Journal on Numerical Analysis.

SIAM Review (Society for Industrial and Applied Mathematics).

USSR Computational Journal of Mathematics and Mathematical Physics.

1.2. ABOUT THE PROGRAMS IN THIS BOOK

The subroutines in this book are a result of years of research and experience of many numerical analysts. We have given considerable thought to deciding which programs to include and which programs to reference in the literature. In addition, we have followed a somewhat uniform style in both the programming and comments. The reader will be able to make better use of these programs if he understands some of our programming conventions. That is the subject of this section.

A major problem encountered by today's scientific programmers is that of *portability* of computer programs. Indeed, one of the motivations for creating high-level languages like Fortran and Algol was the need for programs that will run on different computers. In spite of considerable efforts at standardizing languages like Fortran, many problems of portability remain. Computer manufacturers each have their own brand of Fortran language. Some of these accept standard ANSI Fortran as a subset; some do not. Of the Fortran compilers which accept ANSI Fortran, few give warning messages for statements which are not standard Fortran. The programs in this book were verified to be portable ANSI Fortran using the PFORT Verifier written at Bell Telephone Laboratories by Ryder (1974).

Another major portability problem is the various number representations and arithmetics available on different computers. For example, on the CDC 6600 there are 48 bits of precision in single-precision floating point. The IBM 360 has less than 24. As a result, most floating-point computations performed on the IBM 360 should be done using long precision—nearly 56 bits of precision. But when the same program is used on the CDC 6600, one usually wants to do the computation in single precision. We have therefore supplied both single- and double-precision versions of the programs in this book for actual use. (Only the single-precision versions appear in the text.) To ensure that the two versions of the programs remain identical, the programs have been developed in double precision; conversions to single precision were done automatically using a text editor program.

The job of writing portable programs is further complicated by the fact that many programs make use of machine-dependent constants. The commonest machine-dependent parameter found in mathematical software is the *machine epsilon*. This number, which will be defined precisely in Chapter 2, is related to the accuracy of floating-point arithmetic. Machine-dependent constants such as machine epsilon have traditionally been incorporated into programs as constants. When porting such a program to another machine, someone must find and modify each of the machine-dependent constants. We have avoided this difficulty by computing all such parameters at execution time. These computations require a negligible amount of computer time and storage.

The random number generator URAND given in Chapter 10 automatically determines certain properties of the integer arithmetic system and hence is also portable. However, URAND is available only in single precision, and its precision cannot be changed to double precision.

PROBLEMS

P1-1. The *error function* is an important function in many branches of applied mathematics. It is defined by an integral,

$$\operatorname{erf}(x) = \frac{2}{\sqrt{\pi}} \int_0^x e^{-t^2} \, dt.$$

The integral cannot be expressed in terms of more elementary functions.

Each chapter of this book will include a problem relating the material in the chapter to some aspect of the error function. We begin with

(a) Find a table of erf (x) and look up the value of erf (1.0).

(b) Is there a function or subroutine available on your computer for evaluating erf (x)? If so, find out how to use it, print out the value of erf (1.0), and compare it with the value obtained from the table.

P1-2. Write a short discussion of the various difficulties which may be encountered when converting a Fortran program from double precision to single precision, or vice versa.

P1-3. What is the 9000th digit of π? State precisely how you went about finding it.

P1-4. (F. Acton) This problem offers difficulties that are numerical rather than conceptual. Use a desk or hand calculator with at most ten decimal digits or a pencil and paper, and get as acccurate an answer as you can. Try to get at least five significant decimal digits, and try to estimate the accuracy of your answer. Explain your solution carefully.

A railroad rail exactly one mile long is firmly fixed at both ends and lies in a plane. Some prankster cuts the rail and welds in exactly one additional foot, causing the rail to bow up in an arc of a circle. Determine the maximum height d that this rail now achieves over its former position.

Hints:

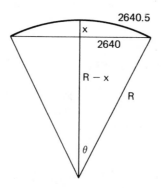

In the diagram, x is the unknown height, R is the unknown radius of the circle, and θ is the angle in radians subtended by half the extended rail. Establish the following relations and use them to compute R, θ, and x:

$$R \sin \theta = 2640$$

$$R\theta = 2640.5$$

$$R \cos \theta = R - x$$

$$\sin \theta = \theta - \frac{\theta^3}{6} + \dots$$

$$\cos \theta = 1 - \frac{\theta^2}{2} + \dots$$

2 FLOATING-POINT COMPUTATION

In this chapter we shall develop some fundamental notions about floating-point computation. We shall first discuss the usual way of approximating the real number system on a computer, i.e., floating-point numbers. The remainder of the chapter is concerned with sources of errors in floating-point computation.

2.1. FLOATING-POINT NUMBERS

The badly named *real number system* underlies the calculus and higher analysis to such an extent that we may forget how impossible it is to represent all real numbers in the real world of finite computers. However, as much as the real number system simplifies analysis, practical computing must do without it.

Many different methods have been proposed for approximating the real number system using finite computer representations. The method used by nearly all computers today is *floating-point numbers*. A set of floating-point numbers F is characterized by four parameters: the number base β, the precision t, and the exponent range $[L, U]$. Each floating-point number x in F has the value

$$x = \pm \left(\frac{d_1}{\beta} + \frac{d_2}{\beta^2} + \ldots + \frac{d_t}{\beta^t} \right) \cdot \beta^e$$

where the integers $d_1, \ldots, d_t$ satisfy

$$0 \le d_i \le \beta - 1 \qquad (i = 1, \ldots, t)$$

and $L \le e \le U$. If for every nonzero x in F, $d_1 \ne 0$, then the floating-point number system F is said to be *normalized*. The integer e is called the *exponent*, and the number $f = (d_1/\beta + \ldots + d_t/\beta^t)$ is called the *fraction*. Usually the integer $\beta^t \cdot f$ is stored using a common integer representation scheme such as signed magnitude, one's complement, or two's complement.

Actual computer implementations of floating-point representations may differ in detail from the ideal ones discussed here, but the differences are minor and can almost always be ignored when dealing with fundamental problems of roundoff errors.

The following table gives some examples of floating point systems. The quantity β^{1-t} is an estimate of the relative accuracy of the arithmetic. We do not give the precise value of machine epsilon because it depends upon complicated details, such as the form of rounding.

If the number of digits, t, is not an integer, it means that $\beta = 2^k$ and $k \cdot t$ bits are available for binary representation of the fraction.

Computer	β	t	L	U	β^{1-t}
Univac 1108	2	27	-128	127	1.49×10^{-8}
Honeywell 6000	2	27	-128	127	1.49×10^{-8}
PDP-11	2	24	-128	127	1.19×10^{-7}
Control Data 6600	2	48	-975	1,070	7.11×10^{-15}
Cray-1	2	48	$-16,384$	8,191	7.11×10^{-15}
Illiac-IV	2	48	$-16,384$	16,383	7.11×10^{-15}
SETUN (Russian)	3	18	?	?	7.74×10^{-9}
Burroughs B5500	8	13	-51	77	1.46×10^{-11}
Hewlett Packard HP-45	10	10	-98	100	1.00×10^{-9}
Texas Instruments SR-5x	10	12	-98	100	1.00×10^{-11}
IBM 360 and 370	16	6	-64	63	9.54×10^{-7}
IBM 360 and 370	16	14	-64	63	2.22×10^{-16}
Telefunken TR440	16	$9\frac{1}{2}$	-127	127	5.84×10^{-11}
Maniac II	65536	$2\frac{11}{16}$	-7	7	7.25×10^{-9}

Some computers use more than one floating-point number system. For example, the IBM 360 uses the two base-16 systems listed above. These two different systems are called *short precision* and *long precision*.

The set F is not a continuum, or even an infinite set. It has exactly $2(\beta - 1)\beta^{t-1}(U - L + 1) + 1$ numbers in it. These are not equally spaced throughout their range but only between successive powers of β. Figure 2.1 shows the 33-point set F for the small illustrative system $\beta = 2, t = 3, L = -1, U = 2$.

Because F is a finite set, there is no possibility of representing the continuum of real numbers in any detail. Indeed, real numbers in absolute value

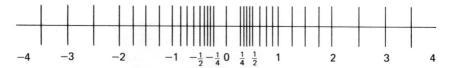

Fig. 2.1. The floating-point number system for $\beta = 2$, $t = 3$, $L = -1$, $U = 2$.

larger than the maximum member of F cannot be said to be represented at all. And, for many purposes, the same is true of nonzero real numbers smaller in magnitude than the smallest positive number in F. Moreover, each number in F has to represent a whole interval of real numbers. If x is a real number in the range of F, we denote by $fl(x)$ a number in F which is closest to x. [Note that if x is equidistant between two numbers in F then $fl(x)$ can take on either of the neighboring values.] It is easy to prove that if x is in the range of F, then

$$\left| \frac{fl(x) - x}{x} \right| \leq \frac{1}{2} \beta^{1-t}.$$

An instructive example is provided by the decimal number 0.1, which is frequently chosen as the step size in many of the algorithms we shall be discussing. Are ten steps of length 0.1 the same as one step of length 1.0? No, not in a floating-point number system for which β is a power of 2, because $\frac{1}{10}$ does not have a terminating expansion in powers of $\frac{1}{2}$. In fact,

$$\frac{1}{10} = \frac{0}{2^1} + \frac{0}{2^2} + \frac{0}{2^3} + \frac{1}{2^4} + \frac{1}{2^5} + \frac{0}{2^6} + \frac{0}{2^7} + \cdots.$$

So, using subscripts to denote the number base,

$$(0.1)_{10} = (0.000110011001100\ldots)_2$$
$$= (0.012121212\ldots)_4$$
$$= (0.063146314\ldots)_8$$
$$= (0.199999999\ldots)_{16}.$$

The quantities on the right must be terminated after t digits. When ten of them are added together, the result is not exactly 1.

As a model of the real number system, the set F has arithmetic operations defined on it, as carried out by the digital computer. Suppose x and y are floating-point numbers. Then the true sum $x + y$ will frequently not be in F. For example, take the 33-point system illustrated in Fig. 2.1; let $x = \frac{5}{4}$ and $y = \frac{3}{8}$. Thus the operation of addition, for example, must itself be simulated on the computer by an approximation called *floating-point addition*. Ideally, if $x, y \in F$ and the number $x + y$ is in the range of F,

$$x \oplus y = fl(x + y),$$

where $\oplus$ denotes floating-point addition. In most computers this ideal is achieved or almost achieved for all such x and y. Thus in our toy 33-point set F we would expect that $\frac{5}{4} \oplus \frac{3}{8}$ would be either $\frac{3}{2}$ or $\frac{7}{4}$. The difference between $x \oplus y$ and $x + y$ (for x, y in F) is the roundoff error made in the floating-point addition. Similar properties hold for the floating-point operations of subtraction, multiplication, and division.

The reason that $\frac{5}{4} + \frac{3}{8}$ is not in the 33-point set F is related to the spacing of the members of F. On the other hand, a sum like $\frac{7}{2} + \frac{7}{2}$ is not in F because 7 is larger than the largest member of F. The attempt to form such a sum on most machines will cause an *overflow signal*, and on most machines the computation will be curtly terminated, for it is considered impossible to provide a useful approximation to numbers beyond the range of F.

While quite a number of the sums $x + y$ (for x, y in F) are themselves in F, it is quite rare for the true product $x \cdot y$ to belong to F, since it will always involve $2t$ or $2t - 1$ significant digits. Moreover, overflow is much more likely in a product. Finally, the phenomenon of *underflow* occurs in floating-point multiplication when two nonzero numbers x, y have a nonzero product that is smaller in magnitude than the smallest nonzero number in F. (Underflow is also possible, though unusual, in addition.) Thus the simulated multiplication operation involves rounding even more often than floating addition.

The operations of floating addition and multiplication are commutative but not associative, and the distributive law also fails for them. Since these algebraic laws are fundamental to mathematical analysis, analyses of floating-point computations are difficult. There are techniques for roundoff error analysis which avoid some of these fundamental difficulties, and many numerical algorithms have been analyzed using these techniques, but they are beyond the scope of this book.

2.2. CALCULATION OF MACHINE EPSILON

The accuracy of floating-point arithmetic can be characterized by *machine epsilon*, the smallest floating-point number ϵ such that

$$1 \oplus \epsilon > 1.$$

Although this definition precisely specifies a unique number in the floating-point number system as machine epsilon, it is not crucial that we use the exact machine epsilon. The value is usually used in a program in such a way that it can differ from its precise definition by a few powers of 2 without causing serious difficulties.

There are several methods for computing the value (or an approximate value) of ϵ. That is, a program can discover the available precision for the machine it is executing on *at execution time*. The method we use for computing an approximation which differs from ϵ by at most a factor of 2 is illustrated by the following segment of a Fortran program:

```
      EPS = 1.
10    EPS = 0.5*EPS
      EPSP1 = EPS + 1.
      IF (EPSP1 .GT. 1.) GO TO 10
```

2.3. AN EXAMPLE OF ROUNDOFF ERROR

An important function of analysis is the exponential function e^x. Since it is used so much, it is essential to be able to compute e^x in a computer program for any floating-point number x. There are, in fact, a great many different methods such an algorithm could use, and most scientific computing systems have one programmed into it. But let us assume such a program does not exist on our computer, and ask how we would program it. This situation might exist for a more obscure transcendental function of analysis or for a new computer.

Recall that, for any real (or even complex) value of x, e^x can be represented as the sum of the convergent infinite series

$$e^x = 1 + x + \frac{x^2}{2!} + \frac{x^3}{3!} + \cdots .$$

We shall try to use this series to compute e^x. Suppose that our floating-point number system is characterized by $\beta = 10$ and $t = 5$. Let us use the series for $x = -5.5$. Here are the numbers we get:

$$
\begin{array}{rr}
e^{-5.5} = & 1.0000 \\
- & 5.5000 \\
+ & 15.125 \\
- & 27.730 \\
+ & 38.129 \\
- & 41.942 \\
+ & 38.446 \\
- & 30.208 \\
+ & 20.768 \\
- & 12.692 \\
+ & 6.9803 \\
- & 3.4902 \\
+ & 1.5997 \\
& \cdot \\
& \cdot \\
& \cdot \\
\hline
+ & 0.0026363
\end{array}
$$

The sum is terminated after 25 terms because subsequent terms no longer change it. Have we computed a satisfactory answer? In fact, $e^{-5.5} = 0.00408677$, so the above series gets an answer with no significant digits.

What went wrong? Notice that some of the terms, and consequently intermediate sums, are several orders of magnitude larger than the final answer. In fact, terms like 38.129 already have roundoff error which is nearly as large as the final result. Indeed, the four leading (i.e., most significant) digits of each of the eight terms that exceed 10 in modulus have all been lost. It would be necessary for these eight terms to be carried to ten significant digits for them each to contribute six significant digits to the answer. Moreover, an eleventh leading digit would be needed to make it likely that the sixth digit would be correct in the sum. This phenomenon is sometimes called *catastrophic cancellation*; it is fairly common in badly conceived computations. However, it is important to realize that this great cancellation is not the cause of error in the answer; it merely magnifies the error already present in the terms.

Although it may be possible to carry more significant digits in a computation to avoid catastrophic cancellation, it is always costly in execution time and storage and often requiries special programming techniques. For this problem there is a much better cure, namely compute the sum for $x = 5.5$, and then take the reciprocal of the answer:

$$e^{-5.5} = \frac{1}{e^{5.5}}$$

$$= \frac{1}{1 + 5.5 + 15.125 + \ldots}$$

$$= 0.0040865, \qquad \text{with our five-decimal arithmetic.}$$

With this computation, the error is reduced to 0.007%.

One of the important points to learn from this example is that a computation does not have to be lengthy to incur serious roundoff errors.

Note how much worse the problem would be if we wanted to compute e^x for $x = -100$.

Actual computer algorithms for calculating e^x start by breaking x into its integer and fractional parts,

$$x = m + f, \qquad \text{integer } m, \ 0 \leq f < 1.$$

Then, using the basic properties of the exponential function,

$$e^x = e^{m+f} = e^m \cdot [e^f]$$

or

$$e^x = e^{(1 + f/m)m} = [e^{1 + f/m}]^m.$$

Either of the quantities in square brackets involves e^y for y on the interval $0 \le y < 2$. Restricting y to such a small range facilitates the computation of e^y. Then the result can be multiplied by e^m (which is easily computed) or raised to the mth power.

2.4. INSTABILITY OF CERTAIN ALGORITHMS

The problem of computing $e^{-5.5}$ discussed in the previous section illustrates how a badly conceived algorithm can get a poor answer to a perfectly well-posed problem. The difficulty was corrected by changing the algorithm. For certain problems, "good" answers cannot be obtained by *any* algorithm because the *problem* is sensitive to the small errors made in data representation and arithmetic. Two examples of such problems are given in Section 2.5. It is important to distinguish between these two classes of pitfalls because there are unstable algorithms and sensitive problems in nearly all branches of numerical mathematics. Once you become aware of their symptoms, these problems are fairly easy to diagnose. In this section we shall discuss a clearer example of an unstable algorithm.

Suppose we wish to compute the integrals

$$E_n = \int_0^1 x^n e^{x-1} \, dx, \qquad n = 1, 2, \ldots .$$

Using integration by parts,

$$\int_0^1 x^n e^{x-1} \, dx = x^n e^{x-1} |_0^1 - \int_0^1 n x^{n-1} e^{x-1} \, dx,$$

or

$$E_n = 1 - n E_{n-1}, \qquad n = 2, \ldots ,$$

where $E_1 = 1/e$. Using $\beta = 10$ and $t = 6$, we can use this recurrence to compute approximations to the first nine successive values of E_n:

$$E_1 \approx 0.367879, \qquad E_6 \approx 0.127120,$$
$$E_2 \approx 0.264242, \qquad E_7 \approx 0.110160,$$
$$E_3 \approx 0.207274, \qquad E_8 \approx 0.118720,$$
$$E_4 \approx 0.170904, \qquad E_9 \approx -0.0684800.$$
$$E_5 \approx 0.145480,$$

Although the integrand $x^9 e^{x-1}$ is positive throughout the interval $(0, 1)$, our computed value for E_9 is negative.

What caused the large error? Observe that the *only* roundoff error made in the above calculations was in E_1 where $1/e$ was rounded to six significant digits. Since the recurrence formula obtained by integration by parts is exact for real arithmetic, the error in E_9 is entirely due to the rounding error in

E_1. To see how the error of about 4.412×10^{-7} in E_1 becomes so large, notice that it is multiplied by -2 in the calculation of E_2, then the error in E_2 is multiplied by -3 in computing E_3, and so on. Thus, the error in E_9 is exactly the error in E_1 multiplied by $(-2)(-3) \ldots (-9) = 9!$ or about $9! \times 4.412 \times 10^{-7} \approx 0.1601$. This enormous magnification of the error in the data of the problem is a result of the algorithm we choose to use. The true value of E_9 (to three significant digits) is $-0.06848 + 0.1601 = 0.0916$.

How can we choose a different algorithm which avoids this instability? If we rewrite the recurrence relation as

$$E_{n-1} = \frac{1 - E_n}{n}, \qquad n = \ldots, 3, 2,$$

then at each stage of the calculation, the error in E_n is decreased by a factor of $1/n$. So if we start with a value for some E_n with $n \gg 1$ and work backwards, any initial error or rounding errors which occur will be decreased at each step. This is called a *stable algorithm*. To obtain a starting value we note that

$$E_n = \int_0^1 x^n e^{x-1}\, dx \leq \int_0^1 x^n\, dx = \frac{x^{n+1}}{n+1}\Big|_0^1 = \frac{1}{n+1}.$$

Thus, E_n goes to zero as n goes to ∞. For example, if we approximate E_{20} by zero and use it as a starting value, we make an initial error of at most $\frac{1}{21}$. This error is multiplied by $\frac{1}{20}$ in computing E_{19}, so the error in E_{19} is at most $(\frac{1}{20})(\frac{1}{21}) \approx 0.0024$. By E_{15}, the initial error has been reduced to less then 4×10^{-8}, which is less than the roundoff error. Doing these calculations gives

$E_{20} \approx 0.0,$	$E_{14} \approx 0.0627322,$
$E_{19} \approx 0.0500000,$	$E_{13} \approx 0.0669477,$
$E_{18} \approx 0.0500000,$	$E_{12} \approx 0.0717733,$
$E_{17} \approx 0.0527778,$	$E_{11} \approx 0.0773523,$
$E_{16} \approx 0.0557190,$	$E_{10} \approx 0.0838771,$
$E_{15} \approx 0.0590176,$	$E_9 \approx 0.0916123.$

By E_{15} the initial error in E_{20} has been completely damped out by the stability of the algorithm, and the computed values of $E_{15}, \ldots, E_9$ are accurate to full six-digit precision, with a possible roundoff error in the last digit.

2.5. SENSITIVITY OF CERTAIN PROBLEMS

We shall now show that certain computational problems are surprisingly sensitive to the data. This aspect of numerical analysis is independent of the floating-point number system or the algorithms used.

A common numerical problem is that of finding the roots of a polynomial; many of these problems are extremely sensitive to the data. For example, consider a quadratic polynomial with nearly multiple roots such as

$$(x - 2)^2 = 10^{-6}.$$

The roots of this equation are 2 ± 10^{-3}. However, a change of 10^{-6} in the constant term can produce a change of 10^{-3} in the roots. This type of instability is even more startling for polynomials of higher degree.

However, it is not only for polynomials with nearly multiple zeros that instability can be observed. The following example is due to Wilkinson (1963). Let

$$p(x) = (x - 1)(x - 2) \ldots (x - 19)(x - 20)$$
$$= x^{20} - 210x^{19} + \ldots .$$

The zeros of $p(x)$ are $1, 2, \ldots , 19, 20$ and are well separated. This example evolved at a place where the floating-point number system had $\beta = 2, t = 30$. To enter a typical coefficient into the computer, it is necessary to round it to 30 significant base-2 digits. Suppose that a change in the 30th significant base-2 digit is made in *only one* of the twenty coefficients. In fact, suppose that the coefficient of x^{19} is changed from -210 to $-210 + 2^{-23}$. How much effect does this small change have on the zeros of the polynomial?

To answer this, Wilkinson carefully computed (using $\beta = 2, t = 90$) the roots of the equation $p(x) + 2^{-23}x^{19} = 0$. These are now listed, correctly rounded to the number of digits shown:

1.00000 0000	10.09526 6145 $\pm$ 0.64350 0904i
2.00000 0000	11.79363 3881 $\pm$ 1.65232 9728i
3.00000 0000	13.99235 8137 $\pm$ 2.51883 0070i
4.00000 0000	16.73073 7466 $\pm$ 2.81262 4894i
4.99999 9928	19.50243 9400 $\pm$ 1.94033 0347i
6.00000 6944	
6.99969 7234	
8.00726 7603	
8.91725 0249	
20.84690 8101	

Note that the small change in the coefficient -210 has caused ten of the zeros to become complex and that two have moved more than 2.81 units off the real axis. Of course, to enter $p(x)$ completely into the computer would require many more roundings, and computing the zeros would cause still more errors.

The above table of zeros was produced by a very accurate computation and does not suffer appreciably from roundoff errors. The reason these zeros moved so far is not a roundoff problem, nor is it a problem related to the algorithm used to compute them; it is a matter of sensitivity of the problem itself. It is easy to analyze what has happened. We can write the polynomial as

$$p(x, \alpha) = x^{20} - \alpha x^{19} + \cdots$$

and then find the partial derivative of x with respect to α at each of the roots of $p(x)$. This is done by differentiating the equation $p(x, \alpha) = 0$ with respect to α:

$$\frac{\partial p(x, \alpha)}{\partial x} \frac{\partial x}{\partial \alpha} + \frac{\partial p(x, \alpha)}{\partial \alpha} = 0,$$

$$\frac{\partial x}{\partial \alpha} = -\frac{\partial p/\partial \alpha}{\partial p/\partial x}$$

$$= \frac{x^{19}}{\sum\limits_{i=1}^{20} \prod\limits_{\substack{j=1 \\ j \neq i}}^{20} (x - j)}.$$

Evaluating this at each root gives

$$\left.\frac{\partial x}{\partial \alpha}\right|_{x=i} = \frac{i^{19}}{\prod\limits_{\substack{j=1 \\ j \neq i}}^{20} (i - j)}, \qquad i = 1, 2, \ldots, 20.$$

These numbers give a direct measure of the sensitivity of each of the roots to the coefficient α. They are listed below:

Root	$\partial x/\partial \alpha\,|_{x=i}$
1	-8.2×10^{-18}
2	8.2×10^{-11}
3	-1.6×10^{-6}
4	2.2×10^{-3}
5	-6.1×10^{-1}
6	5.8×10^{1}
7	-2.5×10^{3}
8	6.0×10^{4}
9	-8.3×10^{5}
10	7.6×10^{6}
11	-4.6×10^{7}
12	2.0×10^{8}
13	-6.1×10^{8}
14	1.3×10^{9}
15	-2.1×10^{9}
16	2.4×10^{9}
17	-1.9×10^{9}
18	1.0×10^{9}
19	-3.1×10^{8}
20	4.3×10^{7}

2.6. SOLVING QUADRATIC EQUATIONS

There is a famous algorithm for solving a quadratic equation, implicit in the following mathematical theorem:

Theorem

If a, b, c are real and $a \neq 0$, then the equation $ax^2 + bx + c = 0$ is satisfied by exactly two values of x, namely

$$x_1 = \frac{-b + \sqrt{b^2 - 4ac}}{2a}$$

and

$$x_2 = \frac{-b - \sqrt{b^2 - 4ac}}{2a}.$$

Let us see how these formulas work when used in a straightforward manner for computing x_1 and x_2. This time we shall use a floating-point system with $\beta = 10$, $t = 8$, $L = -50$, $U = 50$; this has more precision than some widely used computing systems.

Case 1. $a = 1, b = -10^5, c = 1$.

The true roots of the corresponding quadratic equation, correctly rounded to 11 significant decimals, are

$$x_1 = 99999.999990 \qquad \text{(true)},$$
$$x_2 = 0.000010000000001 \qquad \text{(true)}.$$

If we use the expressions of the theorem, we compute

$$x_1 = 100000.00 \qquad \text{(very good)},$$
$$x_2 = 0 \qquad \text{(completely wrong)}.$$

(The reader is advised to be sure he sees how x_2 becomes 0 in this floating-point computation.)

Once again, in computing x_2 we have been a victim of catastrophic cancellation. There are various alternative ways of computing the roots of a quadratic polynomial which avoid such cancellation. One such way is to use the sign of b to determine which of the formulas causes the least cancellation and then use that formula to compute one of the roots. Thus,

$$x_1 = -\frac{b + \text{sign}\,(b)\sqrt{b^2 - 4ac}}{2a}.$$

Then, since $ax^2 + bx + c = a(x - x_1)(x - x_2)$ implies $ax_1x_2 = c$, the other root can be computed by

$$x_2 = \frac{c}{ax_1}.$$

For the problem of Case 1 this method gives $x_1 = 100000.00$, $x_2 = 1.0000000/100000.00 = 0.000010000000$, both acceptable answers.

At this point, we would like to propose the following criterion of performance of a computer algorithm for solving a quadratic equation. This is stated rather loosely here, but a careful statement will be found in Forsythe (1969).

We define a complex number z to be *well within the range of F* if either $z = 0$ or

$$\beta^{L+2} \le \text{Re}\,(z) \le \beta^{U-2} \quad \text{and} \quad \beta^{L+2} \le \text{Im}\,(z) \le \beta^{U-2}.$$

This means that the real and imaginary parts of z are safely within the magnitudes of numbers that can be closely approximated by a member of F. The arbitrary factor β^2 is included to yield a certain margin of safety.

Suppose a, b, c are all numbers in F that are well within the range of F. Then they must be acceptable as input data to the quadratic equation algorithm. If $a = b = c = 0$, the algorithm should terminate with a message signifying that all complex numbers satisfy the equation $ax^2 + bx + c = 0$. If $a = b = 0$ and $c \ne 0$, then the algorithm should terminate with an error message that no complex number satisfies the equation.

Otherwise, let z_1 and z_2 be the exact roots of the equation, so numbered that $|z_1| \le |z_2|$. (If $a = 0$, set $z_2 = \infty$.) Whenever z_1 is well within the range of F, the algorithm should determine an approximation that is close to z_1, in the sense of differing by not more than, say, one unit in the penultimate digit of the root.

The same should be done for z_2.

If either or both of the roots are not well within the range of F, then an appropriate message should be given, and the root (if any) that is well within the range of F should be determined to within a close approximation.

That concludes the loose specification of the desired performance of a quadratic equation-solving algorithm. Let us return to a consideration of some typical equations, to see how the quadratic formulas work with them.

Case 2. $a = 6, b = 5, c = -4$.

There is no difficulty in computing $x_1 = 0.50000000$ and $x_2 = -1.3333333$, or nearly these values, by whatever formula is used.

Case 3. $a = 6 \times 10^{30}, b = 5 \times 10^{30}, c = -4 \times 10^{30}$.

Since these coefficients are those of Case 2, all multiplied by 10^{30}, the roots are unchanged. However, application of the formulas for x_1 and x_2 causes overflow to occur, since $b^2 > 10^{50}$ is out of the range of F. Probably this uniform large size of $|a|, |b|, |c|$ could be detected before entering the algorithm, and all three numbers could be divided through by some scale factor like 10^{30} to reduce the problem to Case 2.

Case 4. $a = 10^{-30}, b = -10^{30}, c = 10^{30}.$

Here x_1 is near 1, while x_2 is near 10^{60}. Thus our algorithm must determine x_1 very closely, even though x_2 is out of the range of F. Obviously any attempt to bring the coefficients to approximate equality in magnitude by simply dividing them all by the same number is doomed to failure and might itself cause an overflow or underflow. This equation is, in fact, a severe test for a quadratic equation solver.

The reader may think that a quadratic equation with one root out of the range of F and one root within the range of F is a contrived example of no practical use. If so, he is mistaken. In many iteration algorithms which solve a quadratic equation as a subroutine, the quadratics have a singular behavior in which $a \longrightarrow 0$ as convergence occurs. One such example is the method given by Muller (1956) for finding zeros of general smooth functions.

Case 5. $a = 1.0000000, b = -4.0000000, c = 3.9999999.$

Here the two roots are $x_1 = 1.999683772$, $x_2 = 2.000316228$. But applying the quadratic formulas gives

$$x_1 = x_2 = 2.0000000,$$

with only the first four digits correct. These roots fail badly to meet the criteria, but the difficulty here is different from that in the other examples.

This quadratic is precisely the quadratic $(x - 2)^2 = e$ with $e = 0.0000001$. As discussed in the previous section, this is a form of sensitivity in the equation itself and not in the method of solving it. However, sensitive quadratic equations can still be solved accurately enough to satisfy our criterion provided additional precision (in this case $t = 16$ or more) is used in part of the calculation.

In a certain sense the roots computed in Case 5 ($x_1 = x_2 = 2$) are "good" solutions of the equation. Notice that these are the exact roots of the equation

$$x^2 - 4x + 4 = 0.$$

This is the same equation as that of Case 5 except for the constant term, which differs by only one in the least significant digit of c. That is, we computed the exact roots for an equation which is very "close" to the equation we are trying to solve.

This last way of looking at rounding errors is called the *inverse error approach* and has been much exploited by J. H. Wilkinson. In general, it is characterized by asking how little change in the data of a problem would be necessary to cause the computed answers to be the exact solution of the changed problem. The more classical way of looking at roundoff, the *direct error approach*, simply asks how wrong the answers are as solutions of the problem with its given data. While both methods are useful, the important feature of the inverse error point of view is that it permits us to analyze round-off error in large matrix or polynomial problems since it permits the use of associative operations, which is often very difficult with the direct error analysis.

Despite the elementary character of the quadratic equation, it is probably still safe to say that (at the time of this writing) not more than five computer programs exist anywhere to meet the criteria outlined above. Creating such an algorithm is not a very deep problem, but it does require attention to the goal and to the details of attaining the goal. It illustrates the sort of place where an undergraduate mathematics or computer science student can make a substantial contribution to computer program libraries.

PROBLEMS

P2-1. The Taylor series for the error function is

$$\text{erf}(x) = \frac{2}{\sqrt{\pi}} \sum_{n=0}^{\infty} \frac{(-1)^n x^{2n+1}}{n!(2n+1)}.$$

The series converges for all x. Write a program to evaluate erf (x) using this series. Use as many terms in the series as are necessary so that the first neglected term does not alter the accumulated sum when it is added to it in floating point. Since this is an alternating series, the error caused by truncating the infinite sum will then be less than the roundoff error. Investigate the effect of roundoff error by comparing the computed sum with the value obtained from a table of erf (x) or the value obtained from a reliable subroutine. Try $x = 0.5, 1.0, 5.0,$ and 10.0. Explain your results.

Hint: The inner loop of your program might look something like this:

```
10   OLDS = S
     EN = EN + 1.
     T = −XSQ*T*(2.*EN−1.)/(EN*(2.*EN+1.))
     S = S + T
     IF (OLDS .NE. S) GO TO 10
```

What values should be assigned to T, S, EN, XSQ before entering this loop? Do not forget the $2/\sqrt{\pi}$ factor.

P2-2.   What output is produced when the following program is run on your computer? Explain why.

```
        X = 0.0
        H = 0.1
        DO 10 I = 1, 10
           X = X + H
   10   CONTINUE
        Y = 1.0 - X
        WRITE(6,20) X, Y
   20   FORMAT(2E20.10)
        STOP
        END
```

P2-3.   Is the decimal number 0.1 exactly representable on your computer? If not, what is the closest floating-point number? What is the floating-point number closest to but less than 0.1? What values are assigned by each of the following statements?

```
        REAL X
        DOUBLE PRECISION Y
        X = 0.1
        Y = 0.1
        Y = 0.1D0
        X = 1.0/10.0
        Y = 1.0D0/10.0D0
        X = 1.0E-1
        Y = 1.0D-1
        READ(5,10) X, Y
   10   FORMAT(E5.1, D5.1)

   Data:  0.1   0.1
```

P2-4.   (a) Is the number $Z = 314,159,265,358$ exactly representable in your computer (1) as an integer? (2) As a single-precision floating-point number? (3) As a double-precision floating-point number?
        (b) What single-precision floating-point number is closest to Z? What value is actually stored by various language systems when Z is assigned to a real variable?
        (c) What double-precision floating-point number is closest to Z? What value is actually stored by various language systems when Z is assigned to a double-precision floating-point variable?
        (d) Find the cosine of 314,159,265,358 radians with an error less than 10^{-9}.
        (e) What happens when you compute cos $(3.14159265358*10^{11})$ in various computer language systems?

P2-5. (Kahan) (a) How are the numbers $\frac{1}{2}$, $\frac{2}{3}$, and $\frac{3}{5}$ represented internally in your computer? Use an appropriate notation, i.e., binary, octal, hexadecimal, etc. How are these numbers represented in the floating-point number systems of other computers such as the IBM 360, CDC 6600, Univac 1108, Honeywell 6000, PDP-11, Burroughs 6500, etc.?

(b) Consider the following Fortran program:

```
        H = 1./2.
        X = 2./3. − H
        Y = 3./5. − H
        E = (X+X+X) − H
        F = (Y+Y+Y+Y+Y) − H
        Q = F/E
        WRITE (6,10) Q
        STOP
   10   FORMAT(1H, G20.10)
        END
```

The variable Q can take on several different values depending on the floating-point arithmetic hardware used by the computer. Try to figure out the value of Q for computers you are familiar with. Run the program on as many computers as you can to check your results. Explain your results.

P2-6. Consider the following two Fortran programs:

```
        EPS = 1.
   10   EPS = EPS/2.
        WRITE (6,20) EPS
   20   FORMAT (1H, G20.10)
        EPSP1 = EPS + 1
        IF (EPSP1 .GT. 1.) GO TO 10
        STOP
        END
```

```
        EPS = 1.
   10   EPS = EPS/2.
        WRITE (6,20) EPS
   20   FORMAT (1H, G20.10)
        IF (EPS .GT. 0.) GO TO 10
        STOP
        END
```

Run the programs on your system, and explain the results.

P2-7. What output is produced when the following Fortran program is run on various computers with which you are familiar? Try to predict the output before actually running the program; then run it to confirm your answer.

```
      EPS = 1.
      KBIT = 0
  10  EPS = .5*EPS
      KBIT = KBIT+1
      EPSP1 = EPS+1.
      IF ((EPSP1.GT.1.).AND.(EPSP1−EPS.EQ.1.)) GO TO 10
      IF (EPSP1−EPS.EQ.1.) EPS = 2.*EPS
      WRITE (6,20) EPS, KBIT
      STOP
  20  FORMAT(1H ,G20.10,I10)
      END
```

What is the effect of including

DOUBLE PRECISION EPS, EPSP1

at the beginning of the program?

P2-8. Consider a three-digit chopped decimal floating-point number system F, with numbers

$$x = \pm.d_1d_2d_3 \times 10^e,$$

with $-100 \le e \le 100$ and $0 \le d_i \le 9$. Let F be normalized (i.e., $d_1 \ne 0$ unless $x = 0$). The number zero is contained in F and has the unique representation $+.000 \times 10^{-100}$.

We denote by $\oplus$ the operation of floating-point addition. For all x and y in F, the value of $x \oplus y$ is defined as the floating-point number closest to $x + y$ whose magnitude is less than or equal to the magnitude of $x + y$. We denote by $\otimes$ the operation of floating-point multiplication. For all x and y in F, the value of $x \otimes y$ is defined as the floating-point number closest to $x \times y$ whose magnitude is less than or equal to the magnitude of $x \times y$.

(a) How many different real numbers can be exactly represented by F?

(b) Find examples of x, y, z in F to show that the following statements are *not* generally true, even if the result is within the range of F:

    (1) $(x \otimes y) \otimes z = x \otimes (y \otimes z)$.

    (2) $(x \oplus y) \oplus z = x \oplus (y \oplus z)$.

(c) Find an example where $(x \oplus y) \oplus z$ has a relative error of at least 50%.

P2-9. Write a subroutine DIVIDE (E, F, A, B, C, D) in double-precision arithmetic that accepts as input four double-precision numbers a, b, c, d and outputs two double-precision numbers e, f such that

$$e + if = \frac{a + ib}{c + id}, \qquad \text{where } i^2 = -1.$$

You may want to include some sort of error parameter for use in case the denominator is zero. Avoid the pitfalls of unnecessary underflows and overflows insofar as you reasonably can. (For example, do not start by computing $c^2 + d^2$, and be sure you can carry out the test case given below. Eliminating *all* pitfalls is too difficult for anyone who is not fairly experienced, but you should explain the remaining weaknesses in your program.)

(a) Test your subroutine with the call
    CALL DIVIDE (E, F, 1.234D38, 2.345D37, 9.876D37, −5.565D36).
(b) For comparison, compute the same quotient using the Fortran built-in division for double-precision complex variables, if available.

P2-10. Write a subroutine SQROOT(E, F, A, B) that uses double-precision real arithmetic and computes one complex square root

$$e + if = \sqrt{a + ib}$$

say the one with $e \geq 0$ and if $e = 0$, then $f \geq 0$. Avoid any pitfalls you reasonably can, and compare your subroutine with the built-in CDSQRT for

$$\sqrt{1.234 \times 10^{38} + i\, 2.345 \times 10^{37}}$$

P2-11. Evaluate the infinite sum

$$\phi(x) = \sum_{k=1}^{\infty} \frac{1}{k(k + x)} \qquad \text{for } x:=0 \text{ step } 0.1 \text{ until } 1.0,$$

with an error less than $0.5*10^{-8}$.

Note: This requires both human analysis and computer power, and neither is likely to succeed without the other. Above all, do not waste a year's computer budget trying to sum the series by brute force. (How much time would you waste?)

Hint: Use the fact that

$$\frac{1}{k(k + 1)} = \frac{1}{k} - \frac{1}{k + 1}$$

to prove that $\phi(1) = 1$. Then express $\phi(x) - \phi(1)$ as an infinite series which converges faster than the one defining $\phi(x)$. You will have to repeat this trick before you get a series for computing $\phi(x)$ that converges fast enough. Reference: Hamming (1962), pp. 48–50.

P2-12. This problem of series evaluation illustrates the impossibility of using modern computer power without analysis in even simple problems.

For $|x| < 1$ we wish to evaluate the expression

$$S(x) = \sum_{k=1}^{\infty} \frac{1}{\sqrt{k^3 + x}} - \sum_{k=1}^{\infty} \frac{1}{\sqrt{k^3 - x}}$$

with an error less than $e = 3 \times 10^{-8}$ in magnitude.
(a) Show that each series converges.
(b) Approximately how many terms would it require to evaluate the first series with an error less than e in magnitude?
(c) Assuming each term can be evaluated in 500 microseconds, how much computer time would it take to evaluate both series to the number of terms you determined under part (b)?
(d) Use some algebra to rewrite the expression $S(x)$ so that it can be evaluated more quickly.
(e) Program your method and evaluate $S(x)$ for two cases: $x = 0.5$ and $x = 0.999999999$. If possible, note the running time and compare it with your estimate.

P2-13. Many times an analytical solution of a problem leads to a series to be evaluated.

Evaluate

$$\sum_{n=1}^{\infty} \frac{1}{n^2 + 1}$$

with an error presumed to be less than 1 in the tenth significant decimal.

Hint:

$$\sum_{n=1}^{\infty} \frac{1}{n^2} = \frac{\pi^2}{6},$$

$$\sum_{n=1}^{\infty} \frac{1}{n^4} = \frac{\pi^4}{90}.$$

Warning: Do not waste machine time summing a series whose rate of convergence you are not sure of.

P2-14. *Research problem*
Write a Fortran subroutine to solve the quadratic equation

$$ax^2 + bx + c = 0,$$

where a, b, c are floating-point numbers of double precision. The algorithm should come as close as possible to the ideals discussed in Section 2.6.

Note: There are two main difficulties:
    1. One must cope with overflow and underflow. This is very serious, mainly because Fortran gives the user program no way to detect underflow and overflow and take remedial

action. The subroutine will have to test whether they are possible and take preventive action first.

2. Even when over- and underflow are impossible, there are cases when satisfactory accuracy cannot be obtained without computing the discriminant $b^2 - 4ac$ with more than the working precision of the arithmetic, for if the root is to be accurate to N significant digits, all the corresponding digits in $\sqrt{b^2 - 4ac}$ must be correct. This may lead you to want to compute $b^2 - 4ac$ in quadruple precision. This will take some special programming. A paper by Dekker (1971) contains some useful programming suggestions.

3 LINEAR SYSTEMS OF EQUATIONS

One of the most frequent problems encountered in scientific computation is the solution of a system of simultaneous linear equations—usually with as many equations as unknowns. Such a system can be written in the form

$$Ax = b,$$

where A is a given square matrix of order n, b is a given column vector of n components, and x is an unknown column vector of n components.

The sources of linear equation problems include the approximation of continuous differential or integral equations by finite, discrete algebraic systems, the local linearization of systems of simultaneous nonlinear equations, and the fitting of polynomials and other curves through data. Some of these applications will be discussed in later chapters.

Students of linear algebra learn methods for solving nonsingular systems of linear equations. One method which is often given in such courses is Cramer's rule, in which each component of the solution is expressed as the quotient of a different numerator determinant over a common denominator determinant. If you tried to solve a system of 20 equations using Cramer's rule, you would need to compute 21 determinants each of order 20. The determinant of a matrix $A = (a_{ij})$ is usually defined as a sum of terms $\pm a_{1-} a_{2-} \ldots a_{n-}$, where the blank subscripts are filled in by some permutation of the integers 1 to n. In this case, the sum has 20! terms, each requiring 19 multiplications. Hence, solution of the linear system takes $21 \times 20! \times 19$ multiplications, plus a similar number of additions. On a fast modern computer, you can perform roughly 100,000 multiplications per second. Thus, the multiplications alone would require about 3×10^8 years provided the computer does not break down part way through the calculations. There are much better ways to compute determinants; however, with a good method, it is possible to solve a linear system in about the time it takes to compute one determinant. Moreover, Cramer's rule often leads to excessive roundoff error.

Students of linear algebra also learn that the solution to $Ax = b$ can be written $x = A^{-1}b$, where A^{-1} is the inverse of A. However, in the vast majority of practical computational problems, it is unnecessary and inadvisable to actually compute A^{-1}. As an extreme but illustrative example, consider a system consisting of just one equation, such as

$$7x = 21.$$

The best way to solve such a system is by division,

$$x = \frac{21}{7} = 3.$$

Use of the matrix inverse would lead to

$$x = (7^{-1})(21)$$
$$= (.142857)(21) = 2.99997.$$

The inverse requires more arithmetic—a division and a multiplication instead of just a division—and produces a less accurate answer. However, the extra arithmetic is the main reason we avoid computing the inverse. Similar considerations apply to systems of more than one equation. This is even true in the common situation where there are several systems of equations involving the same matrix A but different right-hand sides b. Consequently, we shall concentrate on the direct solution of systems of equations rather than the computation of the inverse.

It is important to distinguish between two types of matrices:

1. A *stored matrix* is one for which all n^2 matrix elements a_{ij} are stored in the computer memory. This limits the order n to be less than 100 or so on medium-scale computers and less than several hundred on larger computers.

2. A *sparse matrix* is one for which most of the matrix elements are zero, and the nonzero elements can be either stored in some special data structure or regenerated as needed. (This type of matrix often results from finite-difference and finite-element methods for partial differential equations.) The order n is frequently as large as several tens of thousands and occasionally even larger.

An example of a sparse matrix for which the elements can be easily regenerated is

$$\begin{pmatrix} 4 & 1 & 0 & 0 & 0 & 0 & 0 \\ 1 & 4 & 1 & 0 & 0 & 0 & 0 \\ 0 & 1 & 4 & 1 & 0 & 0 & 0 \\ 0 & 0 & 1 & 4 & 1 & 0 & 0 \\ 0 & 0 & 0 & 1 & 4 & 1 & 0 \\ 0 & 0 & 0 & 0 & 1 & 4 & 1 \\ 0 & 0 & 0 & 0 & 0 & 1 & 4 \end{pmatrix} .$$

These two types overlap somewhat. A stored matrix may have many zero elements and hence also be sparse, but if memory locations are set aside for the zero elements, the sparsity is not important. A very large but non-sparse matrix may be stored in secondary memory such as disk or tape and thereby require more elaborate data-handling techniques. A *band matrix* is one for which all the nonzero elements are near the diagonal, specifically $a_{ij} = 0$ for all i, j with $|i - j| > m$, where $m \ll n$. The *band width* is said to be $2m + 1$, and the nonzero elements are situated on $2m + 1$ diagonals. The matrix shown above is a three-band matrix, called a *tridiagonal matrix*.

Some of the computational techniques appropriate for stored matrices are quite different from those appropriate for sparse matrices. The stored matrix methods are the most basic and will be emphasized here. The methods can be modified to handle matrices in secondary storage, band matrices, and other types of large or moderately sparse matrices.

3.1. LINEAR SYSTEMS FOR STORED MATRICES

In this section we shall consider the solution of a linear system of algebraic equations

$$Ax = b$$

with a stored, *n*-by-*n* matrix A and *n*-vectors b and x. We assume that A is a nonsingular matrix. If A is singular, in principle this will be revealed during the computation, but in practice it may be difficult to decide.

The algorithm that is almost universally used is one of the oldest numerical methods—the systematic elimination method, generally named after C. F. Gauss. Research in the period 1955 to 1965 revealed the importance of two aspects of Gaussian elimination that were not emphasized in earlier work: the search for pivots and the proper interpretation of the effect of rounding errors.

Gaussian elimination and other aspects of matrix computation are studied in detail in the books by Forsythe and Moler (1967) and Stewart (1973).

The reader who wishes more information than we have in this chapter should consult these references.

To illustrate the algorithm, consider the following example of order 3:

$$\begin{pmatrix} 10 & -7 & 0 \\ -3 & 2 & 6 \\ 5 & -1 & 5 \end{pmatrix} \begin{pmatrix} x_1 \\ x_2 \\ x_3 \end{pmatrix} = \begin{pmatrix} 7 \\ 4 \\ 6 \end{pmatrix}.$$

This, of course, represents the three simultaneous equations

$$10x_1 - 7x_2 \qquad = 7,$$
$$-3x_1 + 2x_2 + 6x_3 = 4,$$
$$5x_1 - x_2 + 5x_3 = 6.$$

The first step uses the first equation to eliminate x_1 from the other equations. This is accomplished by adding 0.3 times the first equation to the second equation and adding -0.5 times the first equation to the third equation. The quantities 0.3 and -0.5 are called the *multipliers*.

$$\begin{pmatrix} 10 & -7 & 0 \\ 0 & -0.1 & 6 \\ 0 & 2.5 & 5 \end{pmatrix} \begin{pmatrix} x_1 \\ x_2 \\ x_3 \end{pmatrix} = \begin{pmatrix} 7 \\ 6.1 \\ 2.5 \end{pmatrix}.$$

The second step might involve using the second equation to eliminate x_2 from the third equation. However, the coefficient of x_2 in the second equation is a small number, -0.1. Consequently, the last two equations are interchanged. This is not actually necessary in this example because there are no roundoff errors, but it is crucial in general.

$$\begin{pmatrix} 10 & -7 & 0 \\ 0 & 2.5 & 5 \\ 0 & -0.1 & 6 \end{pmatrix} \begin{pmatrix} x_1 \\ x_2 \\ x_3 \end{pmatrix} = \begin{pmatrix} 7 \\ 2.5 \\ 6.1 \end{pmatrix}.$$

Now, the second equation can be used to eliminate x_2 from the third equation. This is accomplished by adding 0.04 times the second equation to the third equation. (What would the multiplier have been if the equations had not been interchanged?)

$$\begin{pmatrix} 10 & -7 & 0 \\ 0 & 2.5 & 5 \\ 0 & 0 & 6.2 \end{pmatrix} \begin{pmatrix} x_1 \\ x_2 \\ x_3 \end{pmatrix} = \begin{pmatrix} 7 \\ 2.5 \\ 6.2 \end{pmatrix}.$$

The last equation is now

$$6.2x_3 = 6.2.$$

This can be solved to give $x_3 = 1$. This value is substituted into the second equation:

$$2.5x_2 + (5)(1) = 2.5.$$

Hence $x_2 = -1$. Finally the values of x_3 and x_2 are substituted into the first equation:

$$10x_1 + (-7)(-1) = 7.$$

Hence $x_1 = 0$. This solution can be easily checked using the original equations:

$$\begin{pmatrix} 10 & -7 & 0 \\ -3 & 2 & 6 \\ 5 & -1 & 5 \end{pmatrix} \begin{pmatrix} 0 \\ -1 \\ 1 \end{pmatrix} = \begin{pmatrix} 7 \\ 4 \\ 6 \end{pmatrix}.$$

In general, Gaussian elimination involves two stages, the *forward elimination* and the *back substitution*. The forward elimination consists of $n-1$ steps. At the kth step, multiples of the kth equation are subtracted from the remaining equations to eliminate the kth variable. If the coefficient of x_k is "small," it is advisable to interchange equations before this is done. The back substitution consists of solving the last equation for x_n, then the next-to-last equation for x_{n-1}, and so on until x_1 is computed from the first equation.

The entire algorithm can be compactly expressed in matrix notation. For the above example, let

$$M_1 = \begin{pmatrix} 1 & 0 & 0 \\ 0.3 & 1 & 0 \\ -0.5 & 0 & 1 \end{pmatrix}.$$

Then

$$M_1 A = \begin{pmatrix} 10 & -7 & 0 \\ 0 & -0.1 & 6 \\ 0 & 2.5 & 5 \end{pmatrix}, \quad M_1 b = \begin{pmatrix} 7 \\ 6.1 \\ 2.5 \end{pmatrix}.$$

Let

$$P_2 = \begin{pmatrix} 1 & 0 & 0 \\ 0 & 0 & 1 \\ 0 & 1 & 0 \end{pmatrix}, \quad M_2 = \begin{pmatrix} 1 & 0 & 0 \\ 0 & 1 & 0 \\ 0 & 0.04 & 1 \end{pmatrix}.$$

Then

$$M_2 P_2 M_1 A = \begin{pmatrix} 10 & -7 & 0 \\ 0 & 2.5 & 5 \\ 0 & 0 & 6.2 \end{pmatrix}, \quad M_2 P_2 M_1 b = \begin{pmatrix} 7 \\ 2.5 \\ 6.2 \end{pmatrix}.$$

The main point is that because of the way M_1, P_2, and M_2 have been chosen, the product $U = M_2 P_2 M_1 A$ is an *upper triangular matrix*; that is, all its nonzero elements are in the upper right half of the matrix. With such a matrix, the system of equations

$$Ux = c$$

is easily solved by back substitution. By taking $c = M_2 P_2 M_1 b$, the system $Ux = c$ has the same solution as the original system $Ax = b$.

A similar relation is valid in general. Let P_k, $k = 1, \ldots, n-1$, denote the identity matrix, I, modified by interchanging its rows in the same way the rows of A are interchanged at the kth step of the elimination. Let M_k denote the identity matrix modified by inserting the multipliers used at the kth step below the diagonal in the kth column. Each P_k is a *permutation matrix*, and each M_k is a simple lower triangular matrix. Let M and U be the products

$$M = M_{n-1} P_{n-1} \cdot \ldots \cdot M_2 P_2 M_1 P_1,$$

$$U = MA.$$

Then U is an upper triangular matrix, the system $Ux = Mb$ is readily solved for x, and x also solves the original system $Ax = b$.

The matrix M is not necessarily a lower triangular matrix, but it is the product of permutations of simple lower triangular matrices. Consequently the relation $U = MA$ is sometimes called a *triangular decomposition* of A. It should be emphasized that nothing new has been introduced. Triangular factorization is simple Gaussian elimination expressed in matrix notation. [The triangular factors may also be computed by other algorithms; see Forsythe and Moler (1967).]

The diagonal elements of U are called *pivots*. The kth pivot is the coefficient of the kth variable in the kth equation at the kth step of the elimination. In the example, the pivots are 10, 2.5, and 6.2.

Both the computation of the multipliers and the back substitution require divisions by the pivots. Consequently, the algorithm cannot be carried out if any of the pivots are zero. Intuition should tell us that it is a bad idea to complete the computation if any of the pivots are nearly zero. To see this, let us change our example slightly to

$$\begin{pmatrix} 10 & -7 & 0 \\ -3 & 2.099 & 6 \\ 5 & -1 & 5 \end{pmatrix} \begin{pmatrix} x_1 \\ x_2 \\ x_3 \end{pmatrix} = \begin{pmatrix} 7 \\ 3.901 \\ 6 \end{pmatrix}.$$

The $(2, 2)$ element of the matrix has been changed from 2.000 to 2.099, and the right-hand side has also been changed so that the exact answer is still $(0, -1, 1)$. Let us assume that the solution is to be computed on a hypothetical machine which does decimal floating-point arithmetic with five significant digits.

The first step of the elimination produces

$$\begin{pmatrix} 10 & -7 & 0 \\ 0 & -1.0 \times 10^{-3} & 6 \\ 0 & 2.5 & 5 \end{pmatrix} \begin{pmatrix} x_1 \\ x_2 \\ x_3 \end{pmatrix} = \begin{pmatrix} 7 \\ 6.001 \\ 2.5 \end{pmatrix}.$$

The $(2, 2)$ element is now quite small compared with the other elements in the matrix. Nevertheless, let us complete the elimination without using any interchanges. The next step requires adding 2.5×10^3 times the second equation to the third. On the right-hand side, this involves multiplying 6.001 by 2.5×10^3. The result is 1.50025×10^4, which cannot be exactly represented in our hypothetical floating-point number system. It must either be chopped to 1.5002×10^4 or rounded to 1.5003×10^4. The result is then added to 2.5. Let us assume chopped arithmetic is used. Then the last equation becomes

$$1.5005 \times 10^4 x_3 = 1.5004 \times 10^4,$$

and so

$$x_3 = \frac{1.5004 \times 10^4}{1.5005 \times 10^4} = 0.99993.$$

Since the exact answer is $x_3 = 1$, it does not appear that the error is too serious. Unfortunately, x_2 must be determined from the equation

$$-1.0 \times 10^{-3} x_2 + (6)(0.99993) = 6.001,$$

which gives

$$x_2 = \frac{1.5 \times 10^{-3}}{-1.0 \times 10^{-3}} = -1.5.$$

Finally x_1 is determined from the first equation,

$$10x_1 + (-7)(-1.5) = 7,$$

which gives
$$x_1 = -0.35.$$

Instead of $(0, -1, 1)$, we have obtained $(-0.35, -1.50, 0.99993)$.

Where did things go wrong? There was no "accumulation of rounding error" caused by doing thousands of arithmetic operations. The matrix is not close to singular. The difficulty comes from choosing a small pivot at the second step of the elimination. As a result, the multiplier is 2.5×10^3, and the final equation involves coefficients which are 10^3 times as large as those in the original problem. Roundoff errors which are small when compared to these large coefficients are unacceptable in terms of the original matrix and the actual solution.

We leave it to the reader to verify that if the second and third equations are interchanged, then no large multipliers are necessary and the final result is satisfactory. This turns out to be true in general: If the multipliers are all less than or equal to 1 in magnitude, then the computed solution can be proved to be accurate.

Keeping the multipliers less than 1 in absolute value can be ensured by a process known as *partial pivoting*: At the kth step of the forward elimination, the pivot is taken to be the largest (in absolute value) element in the unreduced part of the kth column. The row containing this pivot is interchanged with the kth row to bring the pivot element into the (k, k) position. The same interchanges must be done with the elements of the right-hand side, b. The unknowns in x are not reordered because the columns of A are not interchanged.

If a row or column of A is multiplied by some scale factor prior to the elimination, it is easy to make a compensating change in the computed solution. Such scale factors can be chosen so that no roundoff errors occur during the scaling process, and hence the true solution of the linear system is unchanged. However, scaling the matrix can completely alter the choice of pivots and resulting interchanges and thereby change the accuracy of the computed solution. A common practice is to *equilibrate* the matrix so that the maximum element in each row and column is roughly the same size. However, such a process involves certain difficulties. Consider the matrix

$$A = \begin{pmatrix} 1 & 1 & 1 \\ 10^9 & -1 & 1 \\ 10^9 & 1 & 0 \end{pmatrix}.$$

Dividing the first column of A by 10^9 yields

$$B = \begin{pmatrix} 10^{-9} & 1 & 1 \\ 1 & -1 & 1 \\ 1 & 1 & 0 \end{pmatrix}.$$

On the other hand, dividing the last two rows of A by 10^9 yields

$$C = \begin{pmatrix} 1 & 1 & 1 \\ 1 & -10^{-9} & 10^{-9} \\ 1 & 10^{-9} & 0 \end{pmatrix}.$$

While both B and C are equilibrated, B will be well behaved in a pivoting algorithm, but C will produce a singular submatrix on any computer with less than nine significant decimal digits. (Try it.)

Problems of this type are usually contrived to illustrate a point and rarely occur in practice. When one does occur, the problem or algorithm producing the matrix should be analyzed further before one blindly solves the linear system. The problem of finding a foolproof scaling algorithm is, as yet, unsolved. Until it is, we have decided to accept the original matrix as given and not carry out any scaling.

The rounding errors introduced during the computation almost always cause the computed solution—which we now denote by x_*—to differ somewhat from the theoretical solution, $x = A^{-1}b$. In fact, it *must* differ because the elements of x are usually not floating-point numbers.

There are two common measures of the discrepancy in x_*: the *error*

$$e = x - x_*$$

and the *residual*

$$r = b - Ax_*.$$

Matrix theory tells us that, since A is nonsingular, if one of these is zero, the other must also be zero. But they are not necessarily both "small" at the same time.

Consider the following example:

$$\begin{pmatrix} 0.780 & 0.563 \\ 0.913 & 0.659 \end{pmatrix} \begin{pmatrix} x_1 \\ x_2 \end{pmatrix} = \begin{pmatrix} 0.217 \\ 0.254 \end{pmatrix}.$$

What will happen if we carry out Gaussian elimination with partial pivoting on a hypothetical three-digit decimal computer which truncates? First, the two rows (equations) will be interchanged so that 0.913 becomes the pivot. Then the multiplier

$$\frac{0.780}{0.913} = 0.854 \qquad \text{(to three places)}$$

is computed. Next, 0.854 times the new first row is subtracted from the new second row to produce the system

$$\begin{pmatrix} 0.913 & 0.659 \\ 0 & 0.001 \end{pmatrix} \begin{pmatrix} x_1 \\ x_2 \end{pmatrix} = \begin{pmatrix} 0.254 \\ 0.001 \end{pmatrix}.$$

Finally, the back substitution is carried out:

$$x_2 = \frac{0.001}{0.001} = 1.00 \qquad \text{(exactly)},$$

$$x_1 = \frac{[0.254 - 0.659 x_2]}{0.913}$$

$$= -0.443 \qquad \text{(to three places)}.$$

Thus the computed solution is

$$x_* = \begin{pmatrix} -0.443 \\ 1.000 \end{pmatrix}.$$

To assess the accuracy without knowing the exact answer, we compute the residuals (exactly):

$$r = b - Ax_* = \begin{pmatrix} 0.217 - [(0.780)(-0.443) + (0.563)(1.00)] \\ 0.254 - [(0.913)(-0.443) + (0.659)(1.00)] \end{pmatrix}$$

$$= \begin{pmatrix} -0.000460 \\ -0.000541 \end{pmatrix}.$$

The *residuals* are less than 10^{-3}. We could hardly expect better on a three-digit machine. However, it is easy to see that the exact solution to this system is

$$x = \begin{pmatrix} 1.000 \\ -1.000 \end{pmatrix}.$$

So the *error* is larger than the solution.

    Where the small residuals just a lucky fluke? First, the reader should begin to realize by now that this example is highly contrived. The matrix is incredibly close to being singular and is *not* typical of most problems encountered in practice. Nevertheless, let us track down the reason for the small residuals.

    If Gaussian elimination with partial pivoting is carried out for this example on a computer with *six* or more digits, the forward elimination will produce a system something like

$$\begin{pmatrix} 0.913000 & 0.659000 \\ 0 & 0.000001 \end{pmatrix} \begin{pmatrix} x_1 \\ x_2 \end{pmatrix} = \begin{pmatrix} 0.254000 \\ -0.000001 \end{pmatrix}.$$

Notice that the *sign* of b_2 differs from that obtained with three-digit computation. Now the back substitution produces

$$x_2 = \frac{-0.000001}{0.000001} = -1.00000,$$

$$x_1 = \frac{0.254 - 0.659x_2}{0.913} = 1.00000,$$

the exact answer. On our three-digit machine, x_2 was computed by dividing two quantities both of which were on the order of rounding errors and one of which did not even have the correct sign. Hence x_2 can turn out to be almost anything. (In fact, if we use a machine with nine binary bits, we shall obtain a completely different value.) Then this completely arbitrary value of x_2 was substituted into the first equation to obtain x_1. We can reasonably expect the residual from the first equation to be small—x_1 was computed in such a way as to make this certain. Now comes a subtle but crucial point. We can also expect the residual from the second equation to be small, *precisely because the matrix is so close to being singular.* The two equations are very nearly multiples of one another, so any pair (x_1, x_2) which nearly satisfies the first equation will also nearly satisfy the second. If the matrix were known to be exactly singular, we would not need the second equation at all—any solution of the first would automatically satisfy the second.

Although this example is contrived and atypical, the conclusion we reached is not. It is probably the single most important fact which people concerned with matrix computations have learned in the past 15 or 20 years: *Gaussian elimination with partial pivoting is guaranteed to produce small residuals.*

Now that we have stated it so strongly, we must make a couple of qualifying remarks. By "guaranteed" we mean it is possible to prove a precise theorem which assumes certain technical details about how the floating-point arithmetic system works and which establishes certain inequalities which the components of the residual must satisfy. If the arithmetic units work some other way or if there is a bug in the particular program, then the "guarantee" is void. Furthermore, by "small" we mean on the order of roundoff error *relative to* three things: the elements of the original coefficient matrix, the elements of the coefficient matrix at intermediate steps of the elimination process, and the elements of the computed solution. If any of these are "large," then the residual will not necessarily be small in an absolute sense.

Finally, even if the residual is small, we have made no claims that the error will be small.

The relationship between the size of the residual and the size of the error is determined in part by a quantity known as the *condition number* of the matrix, which is the subject of the next section.

3.2. CONDITION OF A MATRIX

The coefficients in the matrix and right-hand side of a system of simultaneous linear equations are rarely known exactly. Some systems arise from experiments, and so the coefficients are subject to observational errors. Other systems have coefficients given by formulas which involve roundoff error in their evaluation. Even if the system can be stored exactly in the computer, it is almost inevitable that roundoff errors will be introduced during its solution. It can be shown that roundoff errors in Gaussian elimination have the same effect on the answer as errors in the original coefficients.

Consequently, we are led to a fundamental question. If perturbations are made in the coefficients of a system of linear equations, how much is the solution altered? In other words, if $Ax = b$, how can we measure the sensitivity of x to changes in A and b?

The answer to this question lies in making the idea of *nearly singular* precise. If A is a singular matrix, then for some b's a solution x will not exist, while for others it will not be unique. So if A is nearly singular, we can expect small changes in A and b to cause very large changes in x. On the other hand, if A is the identity matrix, then b and x are the same vector. So if A is nearly the identity, small changes in A and b should result in correspondingly small changes in x.

At first glance, it might appear that there is some connection between the size of the pivots encountered in Gaussian elimination with partial pivoting and *nearness to singularity*, because if the arithmetic could be done exactly, all the pivots would be nonzero if and only if the matrix is nonsingular. To some extent, it is also true that if the pivots are small, then the matrix is *close to singular*. However, when roundoff errors are encountered, the converse is no longer true—a matrix might be close to singular even though none of the pivots are small.

To get a more precise, and reliable, measure of nearness to singularity than the size of the pivots, we need to introduce the concept of a *norm* of a vector. This is a single number which measures the general size of the elements of the vector. The most common vector norm is the Euclidean length,

$$\left(\sum_{i=1}^{n} |x_i|^2 \right)^{1/2}.$$

However, use of this norm would make some of our computations too expensive. Instead, in this chapter, we define the norm of a vector with n components to be

$$\|x\| = \sum_{i=1}^{n} |x_i|.$$

This norm has many of the analytic properties of Euclidean length, such as

$$\|x\| > 0 \text{ if } x \neq 0,$$
$$\|0\| = 0,$$
$$\|cx\| = |c| \cdot \|x\| \qquad \text{for all scalars } c,$$
$$\|x + y\| \leq \|x\| + \|y\|.$$

Some of the geometrical properties of Euclidean length are lost, but they are not too important here.

Multiplication of a vector x by a matrix A results in a new vector Ax which may have a very different norm from x. This change in norm is directly related to the sensitivity we wish to measure. The range of the possible change can be expressed by two numbers,

$$M = \max_{x} \frac{\|Ax\|}{\|x\|},$$

$$m = \min_{x} \frac{\|Ax\|}{\|x\|}.$$

The max and min are taken over all nonzero vectors. Note that if A is singular, then $m = 0$. The ratio M/m is called the *condition number* of A,

$$\text{cond}(A) = \frac{\max\limits_{x} \dfrac{\|Ax\|}{\|x\|}}{\min\limits_{x} \dfrac{\|Ax\|}{\|x\|}}.$$

Consider a system of equations

$$Ax = b$$

and a second system obtained by altering the right-hand side:

$$A(x + \Delta x) = b + \Delta b.$$

We think of Δb as being the error in b and Δx as being the resulting error in x, although we need not make any assumptions that the errors are small.

Since $A(\Delta x) = \Delta b$, the definitions of M and m immediately lead to

$$\|b\| \le M \|x\|$$

and

$$\|\Delta b\| \ge m \|\Delta x\|.$$

Consequently, if $m \ne 0$,

$$\frac{\|\Delta x\|}{\|x\|} \le \text{cond}\,(A)\frac{\|\Delta b\|}{\|b\|}.$$

The quantity $\|\Delta b\|/\|b\|$ is the *relative* change in the right-hand side, and the quantity $\|\Delta x\|/\|x\|$ is the *relative* error caused by this change. The advantage of using relative changes is that they are *dimensionless*, that is, they are not affected by overall scale factors.

This shows that the condition number is a relative error magnification factor. Changes in the right-hand side may cause changes cond (A) times as large in the solution. It turns out that the same is true of changes in the coefficient matrix itself.

The condition number is also a measure of nearness to singularity. Although we have not yet developed the mathematical tools necessary to make the idea precise, the condition number can be thought of as the reciprocal of the relative distance from the matrix to the set of singular matrices. So, if cond (A) is large, A is close to singular.

Some of the basic properties of the condition number are easily derived. Clearly, $M \ge m$, and so

$$\text{cond}\,(A) \ge 1.$$

If P is a permutation matrix, then the components of Px are simply a rearrangement of the components of x. It follows that $\|Px\| = \|x\|$ for all x, and so

$$\text{cond}\,(P) = 1.$$

In particular, cond $(I) = 1$. If A is multiplied by a scalar c, then M and m are both multiplied by the same scalar, and so

$$\text{cond}\,(cA) = \text{cond}\,(A).$$

If D is a diagonal matrix, then

$$\text{cond}\,(D) = \frac{\max |d_{ii}|}{\min |d_{ii}|}.$$

The last two properties are two of the reasons that cond (A) is a better measure of nearness to singularity than the determinant of A. As an extreme example, consider a 100-by-100 diagonal matrix with 0.1 on the diagonal. Then det $(A) = 10^{-100}$, which is usually regarded as a small number. But cond $(A) = 1$, and the components of Ax are simply 0.1 times the corresponding components of x. For linear systems of equations, such an A behaves more like the identity than like a singular matrix.

The following example illustrates the condition number.

$$A = \begin{pmatrix} 4.1 & 2.8 \\ 9.7 & 6.6 \end{pmatrix}$$

$$b = \begin{pmatrix} 4.1 \\ 9.7 \end{pmatrix}, \qquad x = \begin{pmatrix} 1 \\ 0 \end{pmatrix}.$$

Clearly, $Ax = b$, and

$$\|b\| = 13.8, \qquad \|x\| = 1.$$

If the right-hand side is changed to

$$b' = \begin{pmatrix} 4.11 \\ 9.70 \end{pmatrix},$$

the solution becomes

$$x' = \begin{pmatrix} 0.34 \\ 0.97 \end{pmatrix}.$$

Let $\Delta b = b - b'$ and $\Delta x = x - x'$. Then

$$\|\Delta b\| = 0.01, \qquad \|\Delta x\| = 1.63.$$

We have made a fairly small perturbation in b which completely changes x. In fact, the relative changes are

$$\frac{\|\Delta b\|}{\|b\|} = 0.0007246, \qquad \frac{\|\Delta x\|}{\|x\|} = 1.63.$$

Since cond (A) is the maximum magnification factor,

$$\text{cond } (A) \geq \frac{1.63}{0.0007246} = 2249.4.$$

We have actually chosen b and Δb so that they give the maximum, and so for this example

$$\text{cond } (A) = 2249.4.$$

It is important to realize that this example is concerned with the *exact* solutions to two slightly different systems of equations and that the method used to obtain the solutions is irrelevant. The example is constructed to have a fairly large condition number so that the effect of changes in b is quite pronounced, but similar behavior can be expected in any problem with a large condition number.

Suppose we wish to solve a problem in which $a_{1,1} = 0.1$, all the other elements of A and b are integers, and cond $(A) = 10^5$. Suppose further that we have a binary computer with 27 bits in the fraction and that we can somehow compute the exact solution to the system actually stored in the computer. Then the only error is caused by representing 0.1 in binary, but we can expect

$$\frac{\|\Delta x\|}{\|x\|} \approx 10^5 \times 2^{-27} \approx 10^{-3}.$$

In other words, the simple act of storing the coefficient matrix in the machine might cause changes in the third significant figures of the true solution.

The condition number also plays a fundamental role in the analysis of the roundoff errors introduced during the solution by Gaussian elimination. Let us assume that A and b have elements which are exact floating-point numbers, and let x_* be the vector of floating-point numbers obtained from a linear equation solver such as the subroutine we shall present in the next section. We also assume that exact singularity is not detected and that there are no underflows or overflows. Then it is possible to establish the following inequalities:

$$\frac{\|b - Ax_*\|}{\|A\|\cdot\|x_*\|} \leq \rho\beta^{-t},$$

$$\frac{\|x - x_*\|}{\|x_*\|} \leq \rho \text{ cond } (A)\beta^{-t}.$$

Here β is the base of the floating-point number system and t is the number of digits, so β^{-t} is about the size of the machine epsilon discussed in Chapter 2. The quantity ρ is defined more carefully later, but it usually has a value no larger than β.

The first inequality says that the *relative residual* can usually be expected to be about the size of roundoff error, no matter how badly conditioned the matrix is. This was illustrated by the example in the previous section. The second inequality requires that A be nonsingular and involves the exact solution x. It follows directly from the first inequality and the definition of cond (A) and says that the *relative error* will also be small if cond (A) is small but might be quite large if the matrix is nearly singular. In the extreme case where A is singular but the singularity is not detected, the first inequality still holds, but the second has no meaning.

To be more precise about the quantity ρ, it is necessary to introduce the idea of a matrix norm and establish some further inequalities. Readers who are not interested in such details can skip the remainder of this section.

The quantity M defined earlier is known as the norm of the matrix. The notation for the matrix norm is the same as for the vector norm,

$$\|A\| = \max_{x} \frac{\|Ax\|}{\|x\|}.$$

Because of our particular definition of $\|x\|$, it is not hard to show that if A has columns a_j, then

$$\|A\| = \max_{j} \|a_j\|.$$

If we had chosen to use the Euclidean length of a vector, then $\|A\|$ would be more expensive to compute. This is discussed in Chapter 9.

The basic result in the study of roundoff error in Gaussian elimination is due to J. H. Wilkinson. He proved that the computed solution x_* exactly satisfies

$$(A + E)x_* = b,$$

where E is a matrix whose elements are about the size of roundoff errors in the elements of A. There are some rare situations where the intermediate matrices obtained during Gaussian elimination have elements which are larger than those of A and there is some effect from accumulation of rounding errors in large matrices, but it can be expected that if ρ is defined by

$$\frac{\|E\|}{\|A\|} = \rho\beta^{-t},$$

then ρ will rarely be bigger than β.

From this basic result, we can immediately derive inequalities involving the residual and the error in the computed solution. The residual is given by

$$b - Ax_* = Ex_*,$$

and hence

$$\|b - Ax_*\| = \|Ex_*\| \leq \|E\|\,\|x_*\|.$$

The residual involves the product Ax_*, so it is appropriate to consider the relative residual which compares the norm of $b - Ax_*$ with the norms of A and x_*. It follows directly from the above inequalities that

$$\frac{\|b - Ax_*\|}{\|A\|\,\|x_*\|} \leq \rho\beta^{-t}.$$

When A is nonsingular, the error can be expressed using the inverse of A by

$$x - x_* = A^{-1}(b - Ax_*),$$

and so

$$\|x - x_*\| \leq \|A^{-1}\| \|E\| \|x_*\|.$$

It is simplest to compare the norm of the error with the norm of the computed solution. Thus the relative error satisfies

$$\frac{\|x - x_*\|}{\|x_*\|} \leq \rho \|A\| \|A^{-1}\| \beta^{-t}.$$

It turns out that $\|A^{-1}\| = 1/m$, and so

$$\text{cond}(A) = \|A\| \|A^{-1}\|.$$

Thus

$$\frac{\|x - x_*\|}{\|x_*\|} \leq \rho \text{ cond}(A)\beta^{-t}.$$

The actual computation of cond (A) involves knowing A^{-1}. If a_j are the columns of A and $\tilde{a}_j$ are the columns of A^{-1}, then in terms of the vector norm we are using

$$\text{cond}(A) = \max_j \|a_j\| \cdot \max_j \|\tilde{a}_j\|.$$

It is easy to compute $\|A\|$, but finding $\|A^{-1}\|$ would roughly triple the time required for Gaussian elimination. Fortunately, the exact value of cond (A) is rarely required. Any reasonably good estimate of it is satisfactory.

The subroutine DECOMP described in the next section estimates the condition of a matrix by

$$\text{cond}(A) \approx \max_j \|a_j\| \frac{\|z\|}{\|y\|},$$

where y and z are two vectors determined by the subroutine so that $\|z\|/\|y\| \approx \|A^{-1}\|$. This involves solving two systems of equations

$$A^T y = e,$$
$$Az = y,$$

where A^T is the transpose of A and e is a vector with components ± 1 chosen to maximize the growth during the back substitution for y.

This estimate is only a lower bound for the actual condition number, but there is some theoretical basis for expecting it to be a fairly accurate estimate.

3.3. SUBROUTINES DECOMP AND SOLVE

Almost any computer library has subroutines based on variants of Gaussian elimination with partial pivoting for solving systems of simultaneous linear equations. The details of implementation of various subroutines available are quite different. These details can have important effects on the execution time of a particular subroutine, but if the subroutine is properly written, they should have little effect on its accuracy.

In this section, we shall describe two such subroutines, DECOMP and SOLVE. DECOMP carries out that part of Gaussian elimination which depends only on the matrix. It saves the multipliers and the pivot information. SOLVE uses these results to obtain the solution for any right-hand side.

DECOMP also returns an estimate of the condition of the matrix. Such an estimate is a much more reliable and useful measure of nearness to singularity than quantities such as the determinant or the smallest pivot.

The estimate is a lower bound for the actual condition, but it is computed in such a way that it is almost always within a factor of n of the actual condition, and it is usually much closer. In other words, for almost all matrices, DECOMP returns a quantity COND with

$$\frac{\text{cond}\,(A)}{n} \leq \text{COND} \leq \text{cond}\,(A).$$

In those situations where $\text{COND} < \text{cond}\,(A)/n$, it still measures the sensitivity of solutions for most right-hand sides.

Roundoff error usually prevents DECOMP, or any other Gaussian elimination subroutine, from determining whether or not the input matrix is singular. If an exact zero pivot occurs during the elimination, DECOMP sets COND to 10^{32} to signal that it has detected singularity. The value 10^{32} is between β^t and β^U on all current floating-point systems, so it is between the reciprocal of the machine accuracy and the overflow level.

However, the occurrence of a zero pivot does not necessarily mean that the matrix is singular, nor does a singular matrix necessarily produce a zero pivot. In fact, the most common source of zero pivots is some kind of bug in the calling program.

It should be realized that, with partial pivoting, *any* matrix has a triangular factorization. DECOMP actually works faster when zero pivots occur because they mean that the corresponding column is already in triangular form. The only difficulty with a zero pivot is that SOLVE will divide by it during the back substitution. So SOLVE should not be used whenever DECOMP has set COND to a value much larger than β^t.

Some of the subroutines available in computer libraries incorporate a technique known as iterative improvement or iterative refinement. This is a process which involves computation of the residual using high-precision arithmetic and solution of a system of equations with the residual as the right-hand side to obtain a correction for the computed solution. The corrected result often has a smaller error but does not necessarily have a smaller residual. Furthermore, the size of the correction is another measure of the sensitivity of the solution to errors in the data and the computation.

We have decided against including an iterative improvement program in this book for several reasons. First, the solution obtained without improvement is satisfactory for most applications. Second, the errors in the input data usually affect the solution more than the roundoff introduced during its computation. Third, our condition estimator supplies the same kind of information available from the size of the correction. Finally, and possibly most important, the availability and use of the required high-precision arithmetic varies from computer to computer. A general linear equation solver which efficiently incorporates iterative improvement cannot be written in standard Fortran. On computers such as the IBM 360 we prefer to use long arithmetic in our normal computation, so the high-precision arithmetic must be done by special subroutines.

To comment upon some details in DECOMP and SOLVE, we need to examine how Fortran systems store matrices. If a program contains the dimension statement

DIMENSION A(3,5),

then $3*5 = 15$ locations will be reserved in memory for the elements of A. They will be stored in the following order:

$$A(1, 1) \quad A(2, 1) \quad A(3, 1) \quad A(1, 2) \quad A(2, 2) \quad A(3, 2) \quad A(1, 3) \ldots.$$

In other words, the elements of each *column* are stored together. The elements of each *row* are separated from each other by a number of locations equal to the first *subscript* in the dimension statement. This convention has been written into the American National Standards Institute specifications for Fortran.

Many of the common matrix operations are most naturally described in terms of rows. For example, in Gaussian elimination, a multiple of one row is subtracted from another row. When implemented in Fortran, such operations typically have the innermost loops varying the second index of arrays. This has two potentially adverse effects on program efficiency. Subscript calculations may be more costly because they involve information contained in the dimension statement. Operating systems which automatically move data between *high-speed* and *secondary* memory units during computation may have to do an excessive amount of work. For these reasons, we have implemented Gaussian elimination in a somewhat unconventional manner with all the inner loops varying the first index. Such an implementation can be significantly more efficient with certain types of operating systems. For details, see Moler (1972).

Most, but not all, Fortran dialects have provision for *variable dimensions* on arrays which are subroutine parameters. In a main program, one may specify

DIMENSION A(50,50)

but intend to actually work with an N-by-N matrix where N may vary from problem to problem. Subroutines such as DECOMP and SOLVE need both N, the actual working order, and the quantity 50 used in the dimension statement because that is the memory increment between successive elements of a row. This dimension information is called NDIM in DECOMP and SOLVE.

DECOMP can be used to compute determinants. This is possible because of three basic properties of determinants. Adding a multiple of one row to another row does not change the determinant. Interchanging two rows changes the sign of the determinant. The determinant of a *triangular* matrix is simply the product of its diagonal elements. DECOMP uses the last component of the pivot vector to return the value $+1$ if an even number of row interchanges is used and the value -1 if an odd number is used. This value can be multiplied by the product of the diagonal elements of the output matrix to obtain the determinant.

One annoying feature of computing determinants is that the intermediate products, and often the determinant itself, tend to be very large or very small numbers and consequently may easily cause floating-point overflow or underflow. One easy remedy is to compute the logarithm of the absolute value of the determinant as the sum of logarithms of its factors.

The following main program illustrates the use of DECOMP and SOLVE. Note that NDIM = 10, the declared dimension of the array A, while $N = 3$, the actual order of the matrix. In our experience with people using these programs for the first time, we have found that improper setting of NDIM is a very frequent source of error.

The right-hand side is set up only after the matrix is found to be sufficiently well conditioned to justify solving a linear system. We have set up the matrix and right-hand side using assignment statements simply to avoid worrying about the format of data.

The example is the one used in Section 3.1. The output is

$$
\begin{array}{ccc}
10. & -7. & 0. \\
-3. & 2. & 6. \\
5. & -1. & 5.
\end{array}
$$

$$COND= \quad 0.12608E+02$$

$$
\begin{array}{c}
7. \\
4. \\
6. \\
0.00000 \\
-1.00000 \\
1.00000
\end{array}
$$

```
C  SAMPLE PROGRAM FOR DECOMP AND SOLVE
C
      REAL A(10,10), B(10), WORK(10), COND, CONDP1
      INTEGER IPVT(10), I, J, N, NDIM
      NDIM = 10
      N = 3
      A(1,1) = 10
      A(2,1) = -3
      A(3,1) = 5
      A(1,2) = -7
      A(2,2) = 2
      A(3,2) = -1
      A(1,3) = 0
      A(2,3) = 6
      A(3,3) = 5
      DO 1 I = 1, N
        WRITE(6,2) (A(I,J), J=1,N)
 1    CONTINUE
 2    FORMAT(1H ,10F5.0)
      WRITE(6,8)
      CALL DECOMP(NDIM, N, A, COND, IPVT, WORK)
      WRITE(6,3) COND
 3    FORMAT(6H COND=, E15.5 )
      WRITE(6,8)
      CONDP1 = COND + 1
      IF (CONDP1 .EQ. COND) WRITE(6,4)
 4    FORMAT(40H MATRIX IS SINGULAR TO WORKING PRECISION)
      IF (CONDP1 .EQ. COND) STOP
      B(1) = 7
      B(2) = 4
      B(3) = 6
      DO 5 I = 1, N
        WRITE(6,2) B(I)
 5    CONTINUE
      WRITE(6,8)
      CALL SOLVE(NDIM, N, A, B, IPVT)
      DO 6 I = 1, N
        WRITE(6,7) B(I)
 6    CONTINUE
 7    FORMAT(1H ,F10.5)
      STOP
 8    FORMAT(1H )
      END
```

```
      SUBROUTINE DECOMP(NDIM,N,A,COND,IPVT,WORK)
C
      INTEGER NDIM,N
      REAL A(NDIM,N),COND,WORK(N)
      INTEGER IPVT(N)
C
C     DECOMPOSES A REAL MATRIX BY GAUSSIAN ELIMINATION
C     AND ESTIMATES THE CONDITION OF THE MATRIX.
C
C     USE SOLVE TO COMPUTE SOLUTIONS TO LINEAR SYSTEMS.
C
C     INPUT..
C
C        NDIM = DECLARED ROW DIMENSION OF THE ARRAY CONTAINING  A.
C
C        N = ORDER OF THE MATRIX.
C
C        A = MATRIX TO BE TRIANGULARIZED.
C
C     OUTPUT..
C
C        A   CONTAINS AN UPPER TRIANGULAR MATRIX   U   AND A PERMUTED
C            VERSION OF A LOWER TRIANGULAR MATRIX  I-L  SO THAT
C            (PERMUTATION MATRIX)*A = L*U
C
C        COND = AN ESTIMATE OF THE CONDITION OF  A .
C            FOR THE LINEAR SYSTEM  A*X = B, CHANGES IN  A   AND  B
C            MAY CAUSE CHANGES  COND  TIMES AS LARGE IN  X .
C            IF  COND+1.0 .EQ. COND , A IS SINGULAR TO WORKING
C            PRECISION.  COND IS SET TO  1.0E+32  IF EXACT
C            SINGULARITY IS DETECTED.
C
C        IPVT = THE PIVOT VECTOR.
C            IPVT(K) = THE INDEX OF THE K-TH PIVOT ROW
C            IPVT(N) = (-1)**(NUMBER OF INTERCHANGES)
C
C     WORK SPACE..  THE VECTOR  WORK  MUST BE DECLARED AND INCLUDED
C                 IN THE CALL.  ITS INPUT CONTENTS ARE IGNORED.
C                 ITS OUTPUT CONTENTS ARE USUALLY UNIMPORTANT.
C
C     THE DETERMINANT OF A CAN BE OBTAINED ON OUTPUT BY
C        DET(A) = IPVT(N) * A(1,1) * A(2,2) * ... * A(N,N).
C
      REAL EK, T, ANORM, YNORM, ZNORM
      INTEGER NM1, I, J, K, KP1, KB, KM1, M
C
      IPVT(N) = 1
      IF (N .EQ. 1) GO TO 80
      NM1 = N - 1
C
C     COMPUTE 1-NORM OF A
C
      ANORM = 0.0
      DO 10 J = 1, N
         T = 0.0
         DO 5 I = 1, N
            T = T + ABS(A(I,J))
    5    CONTINUE
         IF (T .GT. ANORM) ANORM = T
   10 CONTINUE
```

```
C
C     GAUSSIAN ELIMINATION WITH PARTIAL PIVOTING
C
      DO 35 K = 1,NM1
        KP1= K+1
C
C        FIND PIVOT
C
        M = K
        DO 15 I = KP1,N
          IF (ABS(A(I,K)) .GT. ABS(A(M,K))) M = I
   15   CONTINUE
        IPVT(K) = M
        IF (M .NE. K) IPVT(N) = -IPVT(N)
        T = A(M,K)
        A(M,K) = A(K,K)
        A(K,K) = T
C
C        SKIP STEP IF PIVOT IS ZERO
C
        IF (T .EQ. 0.0) GO TO 35
C
C        COMPUTE MULTIPLIERS
C
        DO 20 I = KP1,N
          A(I,K) = -A(I,K)/T
   20   CONTINUE
C
C        INTERCHANGE AND ELIMINATE BY COLUMNS
C
        DO 30 J = KP1,N
          T = A(M,J)
          A(M,J) = A(K,J)
          A(K,J) = T
          IF (T .EQ. 0.0) GO TO 30
          DO 25 I = KP1,N
            A(I,J) = A(I,J) + A(I,K)*T
   25     CONTINUE
   30   CONTINUE
   35 CONTINUE
C
C     COND = (1-NORM OF A)*(AN ESTIMATE OF 1-NORM OF A-INVERSE)
C     ESTIMATE OBTAINED BY ONE STEP OF INVERSE ITERATION FOR THE
C     SMALL SINGULAR VECTOR.  THIS INVOLVES SOLVING TWO SYSTEMS
C     OF EQUATIONS, (A-TRANSPOSE)*Y = E  AND  A*Z = Y  WHERE  E
C     IS A VECTOR OF +1 OR -1 CHOSEN TO CAUSE GROWTH IN Y.
C     ESTIMATE = (1-NORM OF Z)/(1-NORM OF Y)
C
```

```
C     SOLVE (A-TRANSPOSE)*Y = E
C
      DO 50 K = 1, N
        T = 0.0
        IF (K .EQ. 1) GO TO 45
        KM1 = K-1
        DO 40 I = 1, KM1
          T = T + A(I,K)*WORK(I)
 40     CONTINUE
 45     EK = 1.0
        IF (T .LT. 0.0) EK = -1.0
        IF (A(K,K) .EQ. 0.0) GO TO 90
        WORK(K) = -(EK + T)/A(K,K)
 50   CONTINUE
      DO 60 KB = 1, NM1
        K = N - KB
        T = 0.0
        KP1 = K+1
        DO 55 I = KP1, N
          T = T + A(I,K)*WORK(K)
 55     CONTINUE
        WORK(K) = T
        M = IPVT(K)
        IF (M .EQ. K) GO TO 60
        T = WORK(M)
        WORK(M) = WORK(K)
        WORK(K) = T
 60   CONTINUE
C
      YNORM = 0.0
      DO 65 I = 1, N
        YNORM = YNORM + ABS(WORK(I))
 65   CONTINUE
C
C     SOLVE A*Z = Y
C
      CALL SOLVE(NDIM, N, A, WORK, IPVT)
C
      ZNORM = 0.0
      DO 70 I = 1, N
        ZNORM = ZNORM + ABS(WORK(I))
 70   CONTINUE
C
C     ESTIMATE CONDITION
C
      COND = ANORM*ZNORM/YNORM
      IF (COND .LT. 1.0) COND = 1.0
      RETURN
C
C     1-BY-1
C
 80   COND = 1.0
      IF (A(1,1) .NE. 0.0) RETURN
C
C     EXACT SINGULARITY
C
 90   COND = 1.0E+32
      RETURN
      END
```

```
      SUBROUTINE SOLVE(NDIM, N, A, B, IPVT)
C
      INTEGER NDIM, N, IPVT(N)
      REAL A(NDIM,N),B(N)
C
C     SOLUTION OF LINEAR SYSTEM, A*X = B
C     DO NOT USE IF DECOMP HAS DETECTED SINGULARITY
C
C     INPUT..
C
C       NDIM = DECLARED ROW DIMENSION OF ARRAY CONTAINING A
C
C       N = ORDER OF MATRIX.
C
C       A = TRIANGULARIZED MATRIX OBTAINED FROM DECOMP
C
C       B = RIGHT HAND SIDE VECTOR.
C
C       IPVT = PIVOT VECTOR OBTAINED FROM DECOMP
C
C     OUTPUT..
C
C       B = SOLUTION VECTOR, X .
C
      INTEGER KB, KM1, NM1, KP1, I, K, M
      REAL T
C
C     FORWARD ELIMINATION
C
      IF (N .EQ. 1) GO TO 50
      NM1 = N-1
      DO 20 K = 1, NM1
         KP1 = K+1
         M = IPVT(K)
         T = B(M)
         B(M) = B(K)
         B(K) = T
         DO 10 I = KP1, N
            B(I) = B(I) + A(I,K)*T
10       CONTINUE
20    CONTINUE
C
C     BACK SUBSTITUTION
C
      DO 40 KB = 1,NM1
         KM1 = N-KB
         K = KM1+1
         B(K) = B(K)/A(K,K)
         T = -B(K)
         DO 30 I = 1, KM1
            B(I) = B(I) + A(I,K)*T
30       CONTINUE
40    CONTINUE
50    B(1) = B(1)/A(1,1)
      RETURN
      END
```

3.4. LARGE, SPARSE SYSTEMS

Elimination Methods

A number of linear algebraic systems have such a large number, n, of equations and variables that it is impossible to store a full square matrix of n^2 elements. Typically such problems come from the discretization of ordinary or partial differential equations or from problems involving networks or frame structures. Frequently the matrices of these problems are so sparse that there is plenty of high-speed storage for all the nonzero elements, together with some coding which represents the location of each element stored. How shall the associated linear equation system be solved?

When it is possible, Gaussian elimination remains a very economical, accurate, and useful algorithm. Elimination is possible as long as there is space to store all the nonzero elements of the triangular matrices associated with the elimination and when the coding necessary to locate these elements can be programmed. Let LU represent the array whose lower triangle is the matrix of multipliers and whose upper triangle is the triangularized matrix. Then LU is usually more dense than A, although it is still a sparse matrix. The elements of LU that are nonzero in positions where those of A are zero are said to be *filled in* by Gaussian elimination. It is in principle possible to predict the exact extent of the fill-in caused by elimination, but the details may be so complex that the programmer abandons the task.

For certain matrices, the amount of fill-in is easy to bound. One example is a band matrix A. Let us agree for the moment to consider the band matrix as though it were dense, i.e., to ignore any zeros within the band. If Gaussian elimination can be carried out without pivoting, which is safe for certain types of matrices known as *positive definite*; see Forsythe and Moler (1967), then there is no fill-in at all: LU has the same band width as A. If pivoting is necessary (say for an arbitrary nonsingular matrix A), then the fill-in is limited to a band three-halves as wide as A.

It is easy to store the band matrix A in a rectangular array of length n and width $2m + 1$, and the wider band array LU can also be stored and handled easily. As a result, linear systems with band matrices can easily be solved by elimination, provided that the band array LU can be stored in the high-speed storage. A few programs exist for dealing with band matrices; these are comparatively simple modifications of such elimination programs as DECOMP and SOLVE.

If it is necessary to consider the zeros inside the band structure of a matrix, either to save storage or reduce the number of arithmetic operations, then the data structure and its manipulation can become complicated. Such detailed structure may arise with finite-difference methods for a boundary-value problem for an elliptic partial differential equation with variable coefficients.

Dealing with such problems with Gaussian elimination is still a research task, and general-purpose programs are just beginning to appear.

Sometimes a full matrix, or even a band matrix, is too large to keep in the high-speed storage. If so, it is necessary for part of it to be in secondary storage—on the disks or magnetic tape. If a problem is that large, Gaussian elimination is still possible, but the sheer volume of computation makes it rather expensive. The execution time required for the arithmetic operations is usually substantially larger than the time required for transferring parts of the matrix to and from the secondary store. It is therefore important for economy's sake to organize the computation or the operating system environment in such a way that the processor is never waiting for input/output. This can be done in various ways. Programs will not be found on the shelf, although most large installations have had experience with the solution of such large systems.

A good deal of systems effort has recently gone into giving the programmer the illusion of having a very large (so-called *virtual*) high-speed memory available for his data, although in fact the data are grouped into *pages* or *segments* which are constantly being swapped in and out of secondary storage. These systems are found, for example, on the Burroughs B5500 and some IBM 360s and 370s. The presence of the virtual memory keeps the programmer from having to worry about input/output of data. However, this freedom from worry may come at a large price depending on the paging strategy: If the program is forced to wait while the swapping mechanism retrieves each new row of the matrix, as it does with a Burroughs B5500 in batch mode, then the execution time can go up prohibitively for a large matrix.

However, the virtual memory is usually coupled with multiprogramming, and the processor will usually take up another program during the page swap. With many operating systems, one is not charged for this interrupted time, even when another program is not ready. Hence, the "cost" of executing a matrix program remains approximately the same, whether or not there is page swapping. However, whether one is charged or not, swapping prolongs the *elapsed* time until the program is completed.

Iterative Methods

There is a substantial class of linear equation systems for which the elements of A are known by some simple formula and so can be generated as needed. For example, in modeling Laplace's equation with finite-difference schemes, the elements of A are often either 0, 1, or -4, and the values are easily determined by the geometry of a network. Then the elements never need be stored but instead can be generated as needed. Moreover, often the orders n are so large that it would be impossible to store the filled-in array LU.

It is desirable to solve such linear systems $Ax = b$ by methods that never alter the matrix A and never require storing more than a few vectors of length n. (Note that b must usually be stored, as well as x.)

Methods for this purpose exist and are called *iterative*. One starts with a trial solution vector $x^{(0)}$ and carries out some process using A, b, and $x^{(0)}$ to get a new vector $x^{(1)}$. Then one repeats. At the kth stage, one uses the iterative process to get $x^{(k)}$ from A, b, and $x^{(k-1)}$. Under appropriate hypotheses, the vectors $x^{(k)}$ converge to a limit as $k \longrightarrow \infty$. There is a wide variety of such iterative processes. The most successful iterative processes obtain their success from close coupling with the actual problem being solved. Hence it is not usual to find library subroutines for iterative solution methods. Even though the iterative process may be mathematically simple, the structure of the matrix A is likely to be intricate and special to the problem.

One simple iterative method is discussed in Section 24 of Forsythe and Moler (1967): the method known variously as the *Liebmann process*, the *Gauss-Seidel* method, or the *method of successive displacements*. In it, the basic iterative step is to solve the ith equation for the ith component x_i of the new vector x using for each other component of x its most recently computed value. It can be proved to converge for various types of matrices, including any symmetric positive-definite matrix A. However, convergence is ordinarily slow.

With many iterative processes in numerical analysis, convergence is so slow that the most important problem is to find a way of *accelerating* the convergence—e.g., of $x^{(k)}$ to the solution. Indeed, algorithms for accelerating the convergence of sequences form an important part of numerical analysis. The method of *successive overrelaxation* (SOR) is one type of acceleration of the Gauss-Seidel process. It can speed up the convergence to the point where the SOR method is quite widely used in solving finite-difference equations that model elliptic boundary-value problems in two dimensions. Once again it will not be in the program library.

There is a family of iterative methods known as the methods of *conjugate gradients*, or *conjugate directions*. A good explanation of the algorithm can be found in Faddeev and Faddeeva (1963). These methods seem to be relatively successful for symmetric positive-definite matrices and involve no assumptions on the structure of the matrix A. There are a number of published algorithms for the conjugate-gradient method.

PROBLEMS

P3-1. Solve the 3-by-3 system

$$\begin{pmatrix} 1.00 & 0.80 & 0.64 \\ 1.00 & 0.90 & 0.81 \\ 1.00 & 1.10 & 1.21 \end{pmatrix} \begin{pmatrix} x_1 \\ x_2 \\ x_3 \end{pmatrix} = \begin{pmatrix} \text{erf}(0.80) \\ \text{erf}(0.90) \\ \text{erf}(1.10) \end{pmatrix}.$$

See Problem P1-1 for the definition of erf. Use DECOMP and SOLVE. Print out the estimated condition of the matrix and the solution x_1, x_2, x_3. Also print out the sum $x_1 + x_2 + x_3$, and compare it with erf (1.00). Why are the two close to each other? If you cannot answer this last question, see Section 4.1.

P3-2. What is the output from the following program?

```
        DIMENSION A(10,10),WORK(10),IPVT(10)
        NDIM = 10
        N = 2
        A(1,1) = 3.
        A(1,2) = 6.
        A(2,1) = 1.
        A(2,2) = 2.
        CALL DECOMP(NDIM, N, A, COND, IPVT, WORK)
        WRITE(6,1) A(2,2)
        WRITE(6,1) COND
    1   FORMAT(E16.6)
        STOP
        END
```

(a) Explain in detail how the value of $A(2, 2)$ after the call to DECOMP is related to the machine epsilon of your floating-point number system. Explain in general terms how the value of COND is related to machine espilon. (You need not go through all the details of the computation of COND.) If you have access to more than one precision of arithmetic, modify the program, including DECOMP, and answer the same questions for the other precisions.

(b) What output would be produced if the program could be run on the Russian computer, SETUN? (See the table in Section 2.1.)

P3-3. This problem involves verifying two inequalities

$$\frac{\|b - Ax_*\|}{\|A\|\,\|x_*\|} \le \rho \beta^{-t},$$

$$\frac{\|x - x_*\|}{\|x_*\|} \le \rho \,\text{cond}\,(A)\beta^{-t},$$

where the norm of any vector is

$$\|x\| = \sum_{i=1}^{n} |x_i|$$

and the norm of a matrix with columns a_j is

$$\|A\| = \max_j \|a_j\|.$$

You are to experimentally check our claims that ρ in the first inequality is almost always less than β and that the quantity COND returned by DECOMP is a satisfactory substitute for cond (A) in the second inequality.

Choose any matrix A whose elements are exact floating-point numbers. Since the second inequality requires knowing the exact solution, either pick a b for which you know x exactly or pick x and compute $b = Ax$. Make sure there is no roundoff error in A, b, or x so that $Ax = b$ exactly.

Use DECOMP to factor the matrix and compute COND. Use SOLVE to compute x_*. Compute $||A||$, $||x_*||$, $||b - Ax_*||$, and $||x - x_*||$. Be sure to save copies of A and b, since they are altered by the subroutines.

Compute ρ so that the first inequality is actually an equality. If you find that ρ is much larger than β, carefully recheck your program. Large values of ρ are theoretically possible, but they are very rare. They are associated with growth in the size of the elements of the matrix during elimination. See Wilkinson (1963) for more information.

Using your value of ρ, check to see if the second inequality is satisfied with COND in place of cond (A). If it is not, it is because COND is a severe underestimate for the true cond (A). Again, such examples are very rare.

Do this problem with several different matrices, including ones with condition numbers close to 1 and ones with very large condition numbers.

P3-4. The inverse of a matrix A can be defined as the matrix X whose columns x_j satisfy

$$Ax_j = e_j,$$

where e_j is the jth column of the identity matrix. Write a subroutine with the heading

SUBROUTINE INVERT (NDIM, N, A, X, COND, IPVT, WORK)

which accepts a matrix of order N as input and which returns a matrix X, an approximation to the inverse of A, as well as the condition estimate and the pivot information. Your subroutine should call DECOMP just once and call SOLVE a total of N times, once for each column of X. Leave X undefined if DECOMP detects singularity.

Test your subroutine on some matrices whose elements can be exactly represented as floating-point numbers and for which you know A^{-1}. There are several measures of the accuracy of the results:

$$||AX - I||,$$
$$||XA - I||,$$
$$||X - A^{-1}||.$$

You might also use INVERT twice, once to invert A and a second time to invert X. The result is a matrix Z which would be equal to A if there were no roundoff error. So, another measure of accuracy would be

$$||Z - A||.$$

Can you derive an inequality involving ρ, cond (A), and β^{-t} which predicts how large $||Z - A||$ might be?

P3-5. Write a Fortran subroutine that computes the determinant of a square matrix A of order n from the decomposition computed by DECOMP. Use logarithms as suggested in Section 3.3 or some other technique for avoiding underflow and overflow.

P3-6. Let

$$A = \begin{pmatrix} 0.1 & 0.2 & 0.3 \\ 0.4 & 0.5 & 0.6 \\ 0.7 & 0.8 & 0.9 \end{pmatrix}, \quad b = \begin{pmatrix} 0.1 \\ 0.3 \\ 0.5 \end{pmatrix}.$$

(a) Show that the set of linear equations $Ax = b$ has many solutions. Describe the set of possible solutions.

(b) Suppose DECOMP and SOLVE were used to solve $Ax = b$ on a hypothetical computer which does exact arithmetic. Since there are many solutions, it is unreasonable to expect one particular solution to be computed. What does happen?

(c) Use DECOMP and SOLVE to compute a solution on a computer with β a power of 2. Since some of the elements of A are not exact floating-point numbers on such a computer, the matrix which is given to DECOMP is not exactly singular. What solution is obtained? Why? In what sense is it a "good" solution? In what sense is it a "bad" solution?

P3-7. The following tridiagonal matrix occurs in the interpolation of data by cubic splines, as will be seen in the next chapter:

$$A = \begin{pmatrix} -1 & 1 & & & & & \\ 1 & 4 & 1 & & & \mathbf{0} & \\ & 1 & 4 & 1 & & & \\ & & \ddots & \ddots & \ddots & & \\ & & & 1 & 4 & 1 & \\ \mathbf{0} & & & & 1 & 4 & 1 \\ & & & & & 1 & -1 \end{pmatrix}.$$

How does the estimated condition number of this matrix change as its order is increased? What special properties does the factored array returned by DECOMP have? How might the Gaussian elimination algorithm be simplified for this special case? How would you solve a linear system involving this matrix of very high order?

P3-8. It is required to determine the member forces in the 17-member plane truss in the accompanying diagram. The members of the truss are assumed to be joined at the joints by frictionless pins. A theorem of elementary mechanics tells us that since the number of joints j is related to the number of members m by $2j - 3 = m$, the truss is statically determinant. This means that the member forces are determined entirely by the conditions of static equilibrium at the nodes. Let F_x denote horizontal force components and F_y denote

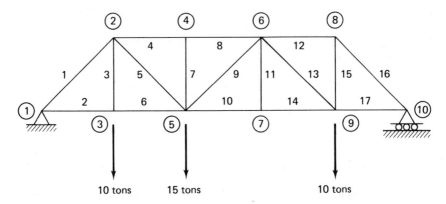

vertical force components. If we let $\alpha = \sin 45° = \cos 45°$ and assume small displacements, then these equilibrium conditions are

$$
\text{joint } 2 \begin{cases} \sum F_x = -\alpha f_1 + f_4 + \alpha f_5 = 0, \\ \sum F_y = -\alpha f_1 - f_3 - \alpha f_5 = 0; \end{cases}
$$

$$
\text{joint } 3 \begin{cases} \sum F_x = -f_2 + f_6 = 0, \\ \sum F_y = f_3 - 10 = 0; \end{cases}
$$

$$
\text{joint } 4 \begin{cases} \sum F_x = -f_4 + f_8 = 0, \\ \sum F_y = -f_7 = 0; \end{cases}
$$

$$
\text{joint } 5 \begin{cases} \sum F_x = -\alpha f_5 - f_6 + \alpha f_9 + f_{10} = 0, \\ \sum F_y = \alpha f_5 + f_7 + \alpha f_9 - 15 = 0; \end{cases}
$$

$$
\text{joint } 6 \begin{cases} \sum F_x = -f_8 - \alpha f_9 + f_{12} + \alpha f_{13} = 0, \\ \sum F_y = -\alpha f_9 - f_{11} - \alpha f_{13} = 0; \end{cases}
$$

$$
\text{joint } 7 \begin{cases} \sum F_x = -f_{10} + f_{14} = 0, \\ \sum F_y = f_{11} = 0; \end{cases}
$$

$$
\text{joint } 8 \begin{cases} \sum F_x = -f_{12} + \alpha f_{16} = 0, \\ \sum F_y = -f_{15} - \alpha f_{16} = 0; \end{cases}
$$

$$
\text{joint } 9 \begin{cases} \sum F_x = -\alpha f_{13} - f_{14} + f_{17} = 0, \\ \sum F_y = \alpha f_{13} + f_{15} - 10 = 0; \end{cases}
$$

$$
\text{joint } 10 \{ \quad \sum F_x = -\alpha f_{16} - f_{17} = 0.
$$

Write a Fortran program which uses DECOMP and SOLVE to solve this linear system of equations for the member forces. Is the matrix of the linear system well conditioned?

4 INTERPOLATION

Suppose one is given a set of real abscissas $x_1, \ldots, x_n$ and corresponding ordinates $y_1, \ldots, y_n$. Here $x_1 < x_2 < \ldots < x_n$, and each y_i is some observed or mathematically defined real number corresponding to x_i. The problem of one-dimensional interpolation is to construct a function f such that $f(x_i) = y_i$ for each i, such that $f(x)$ assumes "reasonable" values for all x between the data points. The criterion of reasonableness may vary from problem to problem and may never be satisfactorily understood.

If the ordinates $\{y_i\}$ come from a smooth mathematical function and they are exact up to a roundoff level, then the problem can be expected to have a satisfactory solution. The reader may be familiar with linear interpolation in a table of logarithms, for example.

If the points (x_i, y_i) come from very precise experimental observations, they often can be considered free from error, and it may be reasonable to interpolate them with a smooth function. If, on the other hand, they arise from relatively crude experiments, it may not be justifiable to force an interpolating function to match the data exactly. By permitting the $f(x_i)$ to differ from y_i, it may be possible to fit the trend of the data very well and even to correct some of their errors. Data fitting that is not exact interpolation is discussed in Chapter 9. Many of the techniques are different.

The purposes of interpolation are many, but they usually amount to the desire to have a fast algorithm to obtain values $f(x)$ for x not in the table of data (x_i, y_i). A short table of data and a short interpolation subroutine may substitute for a very long table of function values. Sometimes one may want to get $f'(x)$ or $f''(x)$ at intermediate points, or one may want to be able to calculate the integral of f over an arbitrary subinterval (a, b) of the interval (x_1, x_n).

The heart of the interpolation problem is a definition of how a reasonable function will behave between data points. After all, the data points can be interpolated by an infinite number of different functions, and we must have some criteria to select among them. The normal criteria are in terms of smoothness and simplicity—e.g., f will be analytic and the maximum value of $|f''(x)|$ throughout the interval will be as small as possible, or f will be a polynomial of minimum degree, or similar conditions.

Most interpolation functions are built out of linear combinations of elementary functions. Linear combinations of the monomial functions $\{x^k\}$ lead to polynomials. Linear combinations of the trigonometric functions $\{\cos kx, \sin kx\}$ lead to trigonometric polynomials. Less often, linear combinations of exponentials $\{\exp(\beta_k x)\}$ or rational functions of the form

$$\frac{\alpha_0 + \alpha_1 x + \ldots + \alpha_m x^m}{\beta_0 + \beta_1 x + \ldots + \beta_n x^n}$$

are also used.

In this chapter we shall deal first with polynomial interpolation and then with a form of piecewise polynomial interpolation called *spline interpolation*.

4.1. POLYNOMINAL INTERPOLATION

For historical as well as pragmatic reasons, the most important class of interpolating functions is the set of algebraic polynomials. Polynomials have the obvious advantage of being easy to evaluate directly. (See Section 4.2.) They can be easily summed, multiplied, integrated, or differentiated. A further property of polynomials is that if c is a constant and $p(x)$ is a polynomial, then so are $p(cx)$ and $p(x + c)$.

Of course, a class of functions can have all of the above properties and still not be satisfactory approximating functions. Fortunately, we have good reason to believe that *any* continuous function $f(x)$ can be closely approximated on a closed interval by some polynomial $p_n(x)$. This follows from an early result of approximation theory known as the *Weierstrass approximation theorem*: *If f is any continuous function on the finite closed interval [a, b], then for every $\epsilon > 0$ there exists a polynomial $p_n(x)$ of degree $n = n(\epsilon)$ such that*

$$\max_{x \in [a, b]} |f(x) - p_n(x)| < \epsilon.$$

The reader is referred to either Ralston (1965) or Wendroff (1966) for a detailed proof.

Although some proofs of Weierstrass' theorem are constructive, the resulting polynomial p_n is generally of such high degree that it is impractical to use. Furthermore, Weierstrass' theorem tells us nothing about the existence of a satisfactory *interpolating* polynomial for a given data set $\{(x_i, y_i)\}$. And while it is comforting to know that some polynomial will approximate $f(x)$ to a specified accuracy throughout the entire interval $[a, b]$, this is no guarantee that such a polynomial will be found by a practical algorithm.

An nth-degree polynomial can be expanded in powers of x,

$$y(x) = a_0 + a_1 x + a_2 x^2 + \ldots + a_n x^n = \sum_{i=0}^{n} a_i x^i,$$

and has $n + 1$ coefficients. Some correctly posed $n + 1$ conditions must be specified to uniquely determine the coefficients. For interpolation we would normally require the polynomial to pass through $n + 1$ points (x_i, y_i), $i = 0$, $1, \ldots, n$, where the x_i are distinct. This gives $n + 1$ linear equations

$$y_i = \sum_{j=0}^{n} a_j x_i^j \qquad (i = 0, 1, \ldots, n)$$

to be solved for the unknowns a_j. We know that any solution is unique by the following argument: Fix the set $\{x_i\}$ of distinct points. If there were two polynomials y and z that both assumed the values y_i at x_i $(i = 0, 1, \ldots, n)$, then the difference polynomial $y - z$ would vanish for $n + 1$ distinct values of x. Since $y - z$ is of degree n, this implies that $y - z$ is the zero polynomial, whence y and z must be the same. This proves the *uniqueness* of the polynomial of degree n interpolating any set of $n + 1$ data points but not its *existence*.

Now the $n + 1$ equations determining the $n + 1$ coefficients of the polynomial y in terms of the data $\{y_i\}$ are linear. One of the major results of linear algebra is that a linear system either has a unique solution for every set of data $\{y_i\}$, or else there are sets of $\{y_i\}$ for which there are many distinct solutions. Since we just proved the impossibility of ever having more than one solution, there must exist a unique polynomial for any interpolation problem with $n + 1$ distinct abscissas.

Thus we may always pass a unique polynomial of degree n through any $n + 1$ given data points having unique abscissas, and this is one approach to the problem of interpolation posed at the beginning of this chapter. Since the entire set of data is fitted by a single polynomial function, it might thus be called a *global* fit of the data.

Once we have decided to fit the data globally by a polynomial of degree n, we are still free to choose whatever basis we want for the space of polynomials of degree n. In the above argument we used the basis $1, x, x^2, \ldots,$ x^n of monomial powers. It leads to a set of linear equations as we saw, and these can in principle be solved by the methods of Chapter 3. However, the equations are surprisingly ill conditioned in many cases. Suppose, for example, that the abscissas $\{x_i\}$ are roughly spaced uniformly within the interval

[0, 1]. Now the successive powers $1, x, x^2, \ldots$ are nearly linearly dependent over the interval [0, 1], partly because they are all positive and their graphs all go from $(0, 0)$ to $(1, 1)$. It is this near linear dependence that makes the linear system quite difficult to solve with normal computing precision for n greater than about 10.

A much more satisfactory way to compute the polynomial that interpolates the points (x_i, y_i) is to use the basis of so-called *Lagrange polynomials* associated with the set $\{x_i\}$. These are polynomials $\{l_j(x)\}$ $(j = 0, 1, \ldots, n)$ of degree n with the property that

$$l_j(x_i) = \begin{cases} 1, & \text{if } i = j, \\ 0, & \text{otherwise.} \end{cases}$$

It is easy to see that the nth-degree polynomial

$$l_j(x) = \frac{(x - x_0)(x - x_1) \ldots (x - x_{j-1})(x - x_{j+1}) \ldots (x - x_n)}{(x_j - x_0)(x_j - x_1) \ldots (x_j - x_{j-1})(x_j - x_{j+1}) \ldots (x_j - x_n)}$$

satisfies these conditions. By our previous work l_j is unique. Each factor in the numerator makes $l_j(x_i)$ zero for some $i \neq j$. Corresponding factors in the denominator normalize the result so that $l_i(x_i) = 1$. The polynomial $l_j(x)y_j$ takes on the value y_j at the point x_j and is zero at all points x_i $(i \neq j)$. Thus, the interpolating polynomial of degree n which passes through the $n + 1$ points (x_i, y_i) is given by

$$y(x) = \sum_{j=0}^{n} l_j(x)y_j.$$

It is clear that the number of arithmetic operations (and consequently the execution time) for this method is proportional to n^2.

It is important to emphasize once more that there is *one and only one* polynomial of degree n or less that interpolates the $n + 1$ data points. There are many different polynomial interpolation formulas in the literature built on different bases; however, for a given set of data values, they all yield the same polynomial. Thus, the polynomial given by this approach is the same as that found by solving the linear equations, provided the calculations are carried out in exact arithmetic.

Roundoff error, storage, and time considerations may influence the choice of methods. A primary consideration in choosing a method is the particular application of the interpolating polynomial. Seldom are the actual coefficients needed. Lagrangian interpolation, or some similar method, is generally a better choice.

It is often best not to use global polynomial interpolation at all. One reason for this point of view is given now: Let an *interpolating array* on an interval $[a, b]$ be any triangular array

$$
\begin{array}{llll}
x_0^1 & & & \\
x_0^2 & x_1^2 & & \\
x_0^3 & x_1^3 & x_2^3 & \\
\cdot & \cdot & \cdot & \\
\cdot & \cdot & \cdot & \cdot \\
\cdot & \cdot & \cdot & \cdot
\end{array}
$$

having the property that all $x_i^j \in [a, b]$ and the elements in any one row are distinct. Let $P_n(f)(x)$ be the interpolating polynomial agreeing with the function $f(x)$ at the points of the $(n + 1)$th row.

Faber's Theorem

For any interpolating array there exists a continuous function g and an x in $[a, b]$ such that $P_n(g)(x)$ does not converge to $g(x)$, as $n \rightarrow \infty$.

In Section 4.3 we shall give an example illustrating Faber's theorem. Problems P4-5 and P4-7 are other convincing examples of the peculiar behavior of polynomial interpolation.

One way to measure the error of interpolation is to suppose that the data points y_i are in fact the values of a known function f at the given abscissas x_i. Suppose that f has $n + 1$ continuous derivatives for all x. Let p_n be the unique polynomial of degree n interpolating the data points $\{(x_i, y_i)\}$. Then it can be proved that, for any real x,

$$
f(x) - p_n(x) = \frac{f^{(n+1)}(\xi)}{(n + 1)!} \prod_{i=0}^{n} (x - x_i),
$$

where ξ is some unknown point in the interval determined by the points $x_0, \ldots, x_n$ and x. When bounds are known for $f^{(n+1)}(x)$, this result can provide a bound for the error.

Many generalizations of Lagrange interpolation exist, the most common of which is *Hermite interpolation*. Here one is given n abscissas $\{x_i\}$, n data values $\{y_i\}$, and n data slopes $\{y_i'\}$. The problem is to find a polynomial $P(x)$ of maximal degree $2n - 1$ such that for $i = 1, 2, \ldots, n$

$$
P(x_i) = y_i
$$

and

$$
P'(x_i) = y_i'
$$

Again if the x_i are distinct, a unique solution exists, and it can be constructed in a manner entirely analogous to that of Lagrange. The reader is referred to Wendroff (1969) or Ralston (1965).

Polynomial interpolation has drawbacks in addition to those of global convergence. The determination and evaluation of interpolating polynomials of high degree can be too time-consuming for certain applications. Polynomials of high degree can also lead to difficult problems associated with round-off error.

4.2. EVALUATION OF POLYNOMIALS

Some programs require repeated evaluation of certain polynomials for various values of their arguments. Thus it is important to be able to evaluate polynomials rapidly.

The evaluation of

$$P(x) = a_1 x^n + a_2 x^{n-1} + \ldots + a_{n+1}$$

by the Fortran code

```
      P = A(N+1)
      DO 10 I = 1, N
         P = P + A(I)*X**(N−I+1)
   10 CONTINUE
```

usually takes $n(n + 1)/2$ multiplications and n additions. A simple technique called *Horner's rule* is achieved by rewriting $P(x)$ as

$$P(x) = a_{n+1} + x(a_n + x(a_{n-1} + x(\ldots(a_2 + a_1 x)\ldots))).$$

This is easily programmed. For example,

```
      P = A(1)
      DO 10 I=1, N
         P = P*X + A(I+1)
   10 CONTINUE
```

Horner's rule takes only n multiplications and n floating-point additions. The same algorithm is also called *synthetic division* or *synthetic substitution* in algebra books.

W. G. Horner's name is attached to this method because he presented the rule in a well-known paper (on another subject) published in 1819. Actually, the rearrangement was published over 100 years earlier by Isaac Newton.

Horner's rule is known to be the *optimal* way to rearrange a polynomial for rapid evaluation without doing substantial computations during the rearrangement. That is, in general, when you are given the coefficients of a polynomial and an argument x, there is no way of evaluating the polynomial using fewer additions and multiplications than with Horner's rule.

For an interesting discussion of polynomial evaluation and some generalizations of Horner's rule, see Knuth (1969).

4.3. AN EXAMPLE, RUNGE'S FUNCTION

The dangers of polynomial interpolation were first discovered by C. Runge in 1901. He attempted to interpolate the simple function

$$y(x) = \frac{1}{1 + 25x^2}$$

in the interval $[-1, 1]$ with polynomials and equally spaced points. He discovered that as the degree n of the interpolating polynomial p_n tends toward infinity, $p_n(x)$ diverges in the intervals $0.726 \ldots \leq |x| < 1$. This phenomenon is shown graphically in Fig. 4.1. Note that global polynomial interpolation worked pretty well for Runge's example in the central portion of the interval.

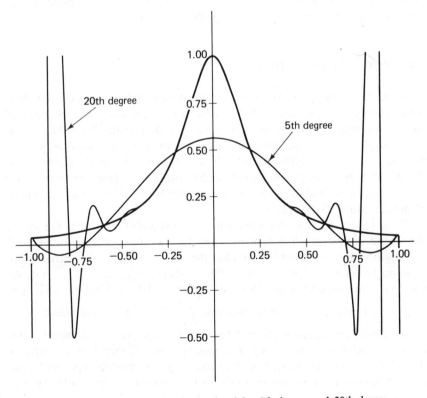

Fig. 4.1. Runge's function interpolated by 5th-degree and 20th-degree interpolating polynomials using equidistant data.

This has led to the concept of fitting data with a *moving polynomial*. For example, one might pass a 10th-degree polynomial through 11 successive data points but use its values only for the central part of this interval. Since spline interpolation works so much better, we shall not discuss moving polynomials further.

If the data abscissas are not equally spaced, but rather placed nearer the ends of the interval at the zeros of the $(n + 1)$th-degree Chebyshev polynomial, the problem with Runge's function disappears. The resulting interpolating polynomial $p_n(x)$ converges to $y(x)$ for every x in $[-1, 1]$ as $n \longrightarrow \infty$.

Of course, the trick of placing the data points at the zeros of a Chebyshev polynomial does not work for all continuous functions. Faber's theorem denies the existence of any such scheme that will work in general.

However, it can be shown that if a function f has a continuous first derivative in $[-1, 1]$, then the interpolating polynomials p_n agreeing with f at the zeros of the $(n + 1)$th-degree Chebyshev polynomials converge to f at every point x in $[-1, 1]$ as $n \rightarrow \infty$. Good references on Chebyshev polynomials are Lanczos (1956), p. 178, and Wendroff (1966), p. 53.

4.4. SPLINE INTERPOLATION

Cubic spline functions are a recent mathematical development, but they model a very old mechanical device. Draftsmen have long used mechanical splines, which are flexible strips of an elastic material, usually wood or (more recently) plastic. The mechanical spline is secured by means of weights at the points of interpolation—historically called *knots*. The spline assumes that shape which minimizes its potential energy, and beam theory states that this energy is proportional to the integral with respect to arc length of the square of the curvature of the spline.

If the spline is a function represented by $s(x)$ and if the slopes are small, the second derivative $s''(x)$ approximates the curvature, and the differential arc length is approximated by dx. Thus the energy of such a *linearized spline* is proportional to $\int s''(x)^2\, dx$. When the knots $(x_1, y_1), (x_2, y_2), \ldots, (x_n, y_n)$ are given, the linearized interpolating spline $s(x)$ is a function such that $s(x_i) = y_i\ (i = 1, 2, \ldots, n)$ and such that $\int_{x_1}^{x_n} (s''(x))^2\, dx$ is minimized.

Assuming the mechanical spline does not break, we would expect s and s' to be continuous on $[x_1, x_n]$. Elementary beam theory further suggests that $s(x)$ is a cubic polynomial between each adjacent pair of knots, and adjacent polynomials join continuously with continuous first and second derivatives.

From a mathematical point of view [see Ahlberg et al. (1967)], the cubic spline function with $s''(x_1) = s''(x_n) = 0$, called the *natural cubic spline*, has been shown to be the unique function possessing the minimum curvature property of all functions interpolating the data and having a square integrable second derivative. In this sense the natural cubic spline is the smoothest function which interpolates the data.

It is worthwhile to count the parameters in a cubic spline curve. In the $n - 1$ intervals between nodes, there are $n - 1$ separate sections of cubic curves, each with four parameters, making $4n - 4$ parameters to be determined. The fact that the function s is continuous and has continuous first and second derivatives at each of the $n - 2$ interior nodes x_i amounts to $3(n - 2)$ conditions on s. Next, the fact that $s(x_i) = y_i$ for each of n nodes imposes n more conditions on s, making the total number of conditions so far equal to $4n - 6$. Thus two more conditions are needed to completely determine the spline. Setting $s''(x_1) = s''(x_n) = 0$ leads to the so-called *natural spline* mentioned above.

Sometimes in curve fitting with cubic splines, instead of the natural end conditions, one imposes some other two conditions at or near the ends of the spline—for example, by prescribing the slopes $s'(x_1)$ and $s'(x_n)$. Such cubic splines minimize the linearized energy integral subject to the two constraints imposed.

The construction of a cubic spline is a simple and numerically stable process. Consider the subinterval $[x_i, x_{i+1}]$ and let

$$h_i = x_{i+1} - x_i,$$

$$w = \frac{x - x_i}{h_i},$$

$$\bar{w} = 1 - w.$$

As x ranges over the subinterval, w goes from 0 to 1 and $\bar{w}$ goes from 1 to 0. Using a little foresight, we decide to represent the spline on this subinterval by

$$s(x) = wy_{i+1} + \bar{w}y_i + h_i^2[(w^3 - w)\sigma_{i+1} + (\bar{w}^3 - \bar{w})\sigma_i],$$

where σ_i and σ_{i+1} are certain constants which are yet to be determined. The first two terms in this expression represent standard linear interpolation, while the term in the brackets is a cubic correction term that will provide the additional smoothness. Notice that the correction term vanishes at the end points so that

$$s(x_i) = y_i$$

and

$$s(x_{i+1}) = y_{i+1}.$$

Thus $s(x)$ interpolates the data no matter how the σ_i are chosen.

We now differentiate $s(x)$ three times, using the chain rule and the fact that $w' = 1/h_i$ and $\bar{w}' = -1/h_i$:

$$s'(x) = \frac{y_{i+1} - y_i}{h_i} + h_i[(3w^2 - 1)\sigma_{i+1} - (3\bar{w}^2 - 1)\sigma_i],$$

$$s''(x) = 6w\sigma_{i+1} + 6\bar{w}\sigma_i,$$

$$s'''(x) = \frac{6(\sigma_{i+1} - \sigma_i)}{h_i}.$$

Note that $s''(x)$ is a linear function which interpolates the values $6\sigma_i$ and $6\sigma_{i+1}$. Consequently,

$$\sigma_i = \frac{s''(x_i)}{6}.$$

This clarifies the meaning of σ_i but does not yet determine its value. Note also that $s'''(x)$ is a constant on each subinterval and that the fourth derivative of $s(x)$ vanishes. This must be true, of course, since $s(x)$ is locally a cubic.

Evaluating $s'(x)$ at the end points of the subinterval gives

$$s'_+(x_i) = \Delta_i - h_i(\sigma_{i+1} + 2\sigma_i),$$

$$s'_-(x_{i+1}) = \Delta_i + h_i(2\sigma_{i+1} + \sigma_i),$$

where $\Delta_i = (y_{i+1} - y_i)/h_i$. We must temporarily use s'_+ and s'_- because our formula for $s(x)$ holds only on $[x_i, x_{i+1}]$, and so the derivatives at the end points are one-sided derivatives. To obtain the desired continuity in $s'(x)$ we impose the following conditions at the interior knots:

$$s'_-(x_i) = s'_+(x_i), \qquad i = 2, \ldots, n - 1.$$

[The continuity of $s''(x)$ then follows immediately from the representation of $s(x)$.] Although the value of $s'_-(x_i)$ comes from considering the subinterval $[x_{i-1}, x_i]$, a formula for it can be obtained simply by replacing i by $i - 1$ in $s'_-(x_{i+1})$. This leads to

$$\Delta_{i-1} + h_{i-1}(2\sigma_i + \sigma_{i-1}) = \Delta_i - h_i(\sigma_{i+1} + 2\sigma_i),$$

and hence

$$h_{i-1}\sigma_{i-1} + 2(h_{i-1} + h_i)\sigma_i + h_i\sigma_{i+1} = \Delta_i - \Delta_{i-1}, \qquad i = 2,\ldots, n - 1.$$

This is a system of $n - 2$ simultaneous linear equations involving n unknowns, $\sigma_i, i = 1, \ldots, n$. Two additional conditions must be specified to uniquely define the interpolating spline. From among the several different ways of picking these two conditions, we have chosen the following.

Let $c_1(x)$ and $c_n(x)$ be the unique cubics which pass through the first four and the last four data points, respectively. The two end conditions match the third derivatives of $s(x)$ to the third derivatives of these cubics, namely

$$s'''(x_1) = c_1'''$$

and

$$s'''(x_n) = c_n'''.$$

The constants c_1''' and c_n''' can be determined directly from the data without actually finding $c_1(x)$ and $c_n(x)$. We have already introduced the quantities

$$\Delta_i = \frac{y_{i+1} - y_i}{x_{i+1} - x_i},$$

which are approximations to first derivatives. Let

$$\Delta_i^{(2)} = \frac{\Delta_{i+1} - \Delta_i}{x_{i+2} - x_i}$$

and

$$\Delta_i^{(3)} = \frac{\Delta_{i+1}^{(2)} - \Delta_i^{(2)}}{x_{i+3} - x_i}.$$

These quantities are known as divided differences; $2\Delta_i^{(2)}$ and $6\Delta_i^{(3)}$ are approximations to second and third derivatives. In particular

$$c_1''' = 6\Delta_1^{(3)}$$

and

$$c_n''' = 6\Delta_{n-3}^{(3)}.$$

Consequently, we require that

$$\frac{\sigma_2 - \sigma_1}{h_1} = \Delta_1^{(3)}$$

and that

$$\frac{\sigma_n - \sigma_{n-1}}{h_{n-1}} = \Delta_{n-3}^{(3)}.$$

To make the final system of equations symmetric, these last two equations should be multiplied by h_1^2 and $-h_{n-1}^2$ to give

$$-h_1\sigma_1 + h_1\sigma_2 = h_1^2\Delta_1^{(3)}$$

and

$$h_{n-1}\sigma_{n-1} - h_{n-1}\sigma_n = -h_{n-1}^2\Delta_{n-3}^{(3)}.$$

For the spline with these end conditions the σ's then satisfy the following system of n linear equations in n unknowns:

$$\begin{pmatrix} -h_1 & h_1 & 0 & 0 & & \\ h_1 & 2(h_1+h_2) & h_2 & 0 & & \mathbf{0} \\ 0 & h_2 & 2(h_2+h_3) & h_3 & & \\ & \cdot & \cdot & \cdot & & \\ & & \cdot & \cdot & \cdot & \\ \mathbf{0} & & \cdot & \cdot & \cdot & \\ & & h_{n-2} & 2(h_{n-2}+h_{n-1}) & h_{n-1} \\ & & 0 & h_{n-1} & -h_{n-1} \end{pmatrix} \begin{pmatrix} \sigma_1 \\ \sigma_2 \\ \sigma_3 \\ \cdot \\ \cdot \\ \cdot \\ \sigma_{n-1} \\ \sigma_n \end{pmatrix} = \begin{pmatrix} h_1^2\Delta_1^{(3)} \\ \Delta_2 - \Delta_1 \\ \Delta_3 - \Delta_2 \\ \cdot \\ \cdot \\ \cdot \\ \Delta_{n-1} - \Delta_{n-2} \\ -h_{n-1}^2\Delta_{n-3}^{(3)} \end{pmatrix}$$

This system of n equations can now be solved by elimination. In fact, DECOMP and SOLVE could be used to compute the σ's. But the matrix of coefficients has several special properties:

1. The matrix is tridiagonal.
2. The matrix is symmetric.
3. For any choice of $x_1 < x_2 < \ldots < x_n$, the matrix is nonsingular and diagonally dominant.

Thus, a unique solution $\sigma_1, \ldots, \sigma_n$ always exists. It can also be shown that for any reasonable choice of $x_1, x_2, \ldots, x_n$ the coefficient matrix is well conditioned. From this fact and the property of diagonal dominance we can expect to calculate accurate solutions using Gaussian elimination without scaling or pivoting.

Applying Gaussian elimination to the original system reduces it to upper triangular form:

$$\begin{pmatrix} \alpha_1 & h_1 & & & & \\ & \alpha_2 & h_2 & & \text{\Large 0} & \\ & & \alpha_3 & h_3 & & \\ & & & \cdot & \cdot & \\ \text{\Large 0} & & & & \cdot & \cdot \\ & & & & & \alpha_n \end{pmatrix} \begin{pmatrix} \sigma_1 \\ \sigma_2 \\ \sigma_3 \\ \cdot \\ \cdot \\ \sigma_n \end{pmatrix} = \begin{pmatrix} \beta_1 \\ \beta_2 \\ \beta_3 \\ \cdot \\ \cdot \\ \beta_n \end{pmatrix},$$

where the diagonal elements α_i are computed by

$$\alpha_1 = -h_1,$$

$$\alpha_i = 2(h_{i-1} + h_i) - \frac{h_{i-1}^2}{\alpha_{i-1}}, \qquad i = 2, 3, \ldots, n-1,$$

$$\alpha_n = -h_{n-1} - \frac{h_{n-1}^2}{\alpha_{n-1}},$$

and the inhomogeneous elements β_i are computed by

$$\beta_1 = h_1^2 \Delta_1^{(3)},$$

$$\beta_i = (\Delta_i - \Delta_{i-1}) - \frac{h_{i-1} \beta_{i-1}}{\alpha_{i-1}}, \qquad i = 2, 3, \ldots, n-1.$$

$$\beta_n = -h_{n-1}^2 \Delta_{n-3}^{(3)} - h_{n-1} \frac{\beta_{n-1}}{\alpha_{n-1}}.$$

Finally, the coefficients σ_i are computed by the back substitution

$$\sigma_n = \frac{\beta_n}{\alpha_n},$$

$$\sigma_i = \frac{\beta_i - h_i \sigma_{i+1}}{\alpha_i}, \qquad i = n-1, n-2, \ldots, 1.$$

Depending on the application, one may wish to store the σ_i and use them directly to evaluate the spline. If the spline is to be evaluated many times, it is worthwhile to rearrange terms. For various reasons, it is preferable to calculate and store the actual cubic coefficients b_i, c_i, and d_i, $i = 1, 2, \ldots$, $n - 1$, for each interval $[x_i, x_{i+1}]$, where

$$s(x) = y_i + b_i(x - x_i) + c_i(x - x_i)^2 + d_i(x - x_i)^3, \qquad x_i \leq x \leq x_{i+1}.$$

These coefficients are given by

$$b_i = \frac{y_{i+1} - y_i}{h_i} - h_i(\sigma_{i+1} + 2\sigma_i),$$

$$c_i = 3\sigma_i,$$

$$d_i = \frac{\sigma_{i+1} - \sigma_i}{h_i}$$

for $i = 1, 2, \ldots, n - 1$. Using this form for storing the spline simplifies manipulations such as derivatives and integrals.

4.5. SUBROUTINES SPLINE AND SEVAL

The method of calculating the parameters of a cubic spline function that was discussed in the previous section is implemented in the subroutine SPLINE. The comments in the program describe how the subroutine is used. During the elimination, the arrays B, C, and D are used to store the quantities α_i, β_i, and h_i of the previous section.

The function subprogram SEVAL can be used to evaluate the spline function once the coefficients have been calculated by SPLINE. If the independent variable U is not in the same interval as in the previous call, a binary search is employed for determining the proper interval. If the reader is not familiar with how a binary search works, he is urged to try a few examples by hand. A binary search is certainly not the best technique for *all* applications (why?).

A simple example follows that shows the use of SPLINE and SEVAL. The output is 2.50000 15.62500. (Why?)

```
C  SAMPLE PROGRAM FOR SPLINE AND SEVAL
C
       REAL X(10), Y(10), B(10), C(10), D(10)
       REAL S, U, SEVAL
       INTEGER  I, N
C
       N = 10
       DO 1 I = 1,N
         X(I) = I
         Y(I) = X(I)**3
     1 CONTINUE
C
       CALL SPLINE ( N, X, Y, B, C, D)
C
       U = 2.5
       S = SEVAL(N, U, X, Y, B, C, D)
       WRITE(6,2) U, S
     2 FORMAT(2F10.5)
       STOP
       END
```

```
      SUBROUTINE SPLINE (N, X, Y, B, C, D)
      INTEGER N
      REAL X(N), Y(N), B(N), C(N), D(N)
C
C  THE COEFFICIENTS B(I), C(I), AND D(I), I=1,2,...,N ARE COMPUTED
C  FOR A CUBIC INTERPOLATING SPLINE
C
C     S(X) = Y(I) + B(I)*(X-X(I)) + C(I)*(X-X(I))**2 + D(I)*(X-X(I))**3
C
C     FOR   X(I) .LE. X .LE. X(I+1)
C
C  INPUT..
C
C    N = THE NUMBER OF DATA POINTS OR KNOTS (N.GE.2)
C    X = THE ABSCISSAS OF THE KNOTS IN STRICTLY INCREASING ORDER
C    Y = THE ORDINATES OF THE KNOTS
C
C  OUTPUT..
C
C    B, C, D  = ARRAYS OF SPLINE COEFFICIENTS AS DEFINED ABOVE.
C
C  USING  P  TO DENOTE DIFFERENTIATION,
C
C    Y(I) = S(X(I))
C    B(I) = SP(X(I))
C    C(I) = SPP(X(I))/2
C    D(I) = SPPP(X(I))/6  (DERIVATIVE FROM THE RIGHT)
C
C  THE ACCOMPANYING FUNCTION SUBPROGRAM  SEVAL  CAN BE USED
C  TO EVALUATE THE SPLINE.
C
C
      INTEGER NM1, IB, I
      REAL T
C
      NM1 = N-1
      IF ( N .LT. 2 ) RETURN
      IF ( N .LT. 3 ) GO TO 50
C
C  SET UP TRIDIAGONAL SYSTEM
C
C  B = DIAGONAL, D = OFFDIAGONAL, C = RIGHT HAND SIDE.
C
      D(1) = X(2) - X(1)
      C(2) = (Y(2) - Y(1))/D(1)
      DO 10 I = 2, NM1
         D(I) = X(I+1) - X(I)
         B(I) = 2.*(D(I-1) + D(I))
         C(I+1) = (Y(I+1) - Y(I))/D(I)
         C(I) = C(I+1) - C(I)
   10 CONTINUE
```

```
C
C   END CONDITIONS.  THIRD DERIVATIVES AT  X(1)  AND  X(N)
C   OBTAINED FROM DIVIDED DIFFERENCES
C
      B(1) = -D(1)
      B(N) = -D(N-1)
      C(1) = 0.
      C(N) = 0.
      IF ( N .EQ. 3 ) GO TO 15
      C(1) = C(3)/(X(4)-X(2)) - C(2)/(X(3)-X(1))
      C(N) = C(N-1)/(X(N)-X(N-2)) - C(N-2)/(X(N-1)-X(N-3))
      C(1) = C(1)*D(1)**2/(X(4)-X(1))
      C(N) = -C(N)*D(N-1)**2/(X(N)-X(N-3))
C
C   FORWARD ELIMINATION
C
   15 DO 20 I = 2, N
      T = D(I-1)/B(I-1)
      B(I) = B(I) - T*D(I-1)
      C(I) = C(I) - T*C(I-1)
   20 CONTINUE
C
C   BACK SUBSTITUTION
C
      C(N) = C(N)/B(N)
      DO 30 IB = 1, NM1
      I = N-IB
      C(I) = (C(I) - D(I)*C(I+1))/B(I)
   30 CONTINUE
C
C   C(I) IS NOW THE SIGMA(I) OF THE TEXT
C
C   COMPUTE POLYNOMIAL COEFFICIENTS
C
      B(N) = (Y(N) - Y(NM1))/D(NM1) + D(NM1)*(C(NM1) + 2.*C(N))
      DO 40 I = 1, NM1
      B(I) = (Y(I+1) - Y(I))/D(I) - D(I)*(C(I+1) + 2.*C(I))
      D(I) = (C(I+1) - C(I))/D(I)
      C(I) = 3.*C(I)
   40 CONTINUE
      C(N) = 3.*C(N)
      D(N) = D(N-1)
      RETURN
C
   50 B(1) = (Y(2)-Y(1))/(X(2)-X(1))
      C(1) = 0.
      D(1) = 0.
      B(2) = B(1)
      C(2) = 0.
      D(2) = 0.
      RETURN
      END
```

```
      REAL FUNCTION SEVAL(N, U, X, Y, B, C, D)
      INTEGER N
      REAL U, X(N), Y(N), B(N), C(N), D(N)
C
C THIS SUBROUTINE EVALUATES THE CUBIC SPLINE FUNCTION
C
C   SEVAL = Y(I) + B(I)*(U-X(I)) + C(I)*(U-X(I))**2 + D(I)*(U-X(I))**3
C
C   WHERE  X(I) .LT. U .LT. X(I+1), USING HORNER'S RULE
C
C IF  U .LT. X(1) THEN  I = 1  IS USED.
C IF  U .GE. X(N) THEN  I = N  IS USED.
C
C INPUT..
C
C   N = THE NUMBER OF DATA POINTS
C   U = THE ABSCISSA AT WHICH THE SPLINE IS TO BE EVALUATED
C   X,Y = THE ARRAYS OF DATA ABSCISSAS AND ORDINATES
C   B,C,D = ARRAYS OF SPLINE COEFFICIENTS COMPUTED BY SPLINE
C
C IF  U  IS NOT IN THE SAME INTERVAL AS THE PREVIOUS CALL, THEN A
C BINARY SEARCH IS PERFORMED TO DETERMINE THE PROPER INTERVAL.
C
      INTEGER I, J, K
      REAL DX
      DATA I/1/
      IF ( I .GE. N ) I = 1
      IF ( U .LT. X(I) ) GO TO 10
      IF ( U .LE. X(I+1) ) GO TO 30
C
C BINARY SEARCH
C
   10 I = 1
      J = N+1
   20 K = (I+J)/2
      IF ( U .LT. X(K) ) J = K
      IF ( U .GE. X(K) ) I = K
      IF ( J .GT. I+1 ) GO TO 20
C
C EVALUATE SPLINE
C
   30 DX = U - X(I)
      SEVAL = Y(I) + DX*(B(I) + DX*(C(I) + DX*D(I)))
      RETURN
      END
```

PROBLEMS

P4-1. Generate $n = 11$ data points by taking

$$\left.\begin{array}{l} x_i = \dfrac{i - 1}{10} \\[2mm] y_i = \text{erf}\,(x_i) \end{array}\right\} \quad i = 1, \ldots, n.$$

Obtain the values of erf from published tables or a subroutine on your system. Use SPLINE to find the coefficients for the cubic spline interpolating the data. Use SEVAL to evaluate the spline at points between the data points and compare the value of the spline with the value of erf. What is the maximum error you find at these intermediate points?

P4-2. Use the subroutines SPLINE and SEVAL to interpolate Runge's function

$$f(x) = \frac{1}{1 + 25x^2}$$

at the $n = 21$ points $x_i = -1.0, -0.9, \ldots, 0.9, 1.0$. Compare the results with the 20th-degree polynomial which interpolates the same data.

P4-3. Modify SPLINE so that it finds the coefficients of the *natural* cubic spline, that is, the spline with end conditions $s''(x_1) = s''(x_n) = 0$. The modifications will be fairly extensive because, among other things, there are now only $n - 2$ unknown σ's. Is it also necessary to modify SEVAL?

P4-4. Compare the natural spline and the spline with the end conditions we have described. Specifically, generate test data by taking x_i, $i = 1, \ldots, n$, to be n equally spaced points in the interval $[0, \pi]$, including the end points, and $y_i = \cos(x_i)$. Obtain the spline coefficients from the subroutine SPLINE without any modifications. Also obtain the coefficients from SPLINE as modified in Problem P4-3. Compare the two different functions $s(x)$ with $\cos(x)$ for values of x other than the data points. Pay particular attention to values near the ends of the interval. Use several values of n, say 10, 20, and 30. Why should $\cos(x)$ be used instead of $\sin(x)$ in this experiment?

P4-5. The following figures from the Census Bureau give the population of the United States:

Year	Population
1900	75,994,575
1910	91,972,266
1920	105,710,620
1930	122,775,046
1940	131,669,275
1950	150,697,361
1960	179,323,175
1970	203,235,298

(a) Since there are eight points, there is a unique polynomial of degree 7 which interpolates the data. However, some of the ways of representing this polynomial are computationally more satisfactory than others. Here are four possibilities, each with t ranging over the interval $1900 \le t \le 1970$:

$$\sum_{j=0}^{7} a_j t^j,$$

$$\sum_{j=0}^{7} b_j (t - 1900)^j,$$

$$\sum_{j=0}^{7} c_j (t - 1935)^j,$$

$$\sum_{j=0}^{7} d_j \left(\frac{t - 1935}{35}\right)^j.$$

In each case, the coefficients are found by solving an 8-by-8 system of equations, but the matrices of the various systems are quite different. Set up each of the four matrices, and find the estimate of its condition using DECOMP. Then use SOLVE to find the coefficients. Check each of the representations to see how well it reproduces the original data.

(b) Interpolate the data by a 7th-degree polynomial, using the best conditioned representation found in part (a), and by a cubic spline using SPLINE. Tabulate and graph the resulting functions at one-year intervals over the period from 1900 to 1980. You should find that the two functions agree fairly well up to 1970 but that their behavior between 1970 and 1980 is quite different. Which of the two approaches predicts *zero population growth* before 1980? If you are doing this problem after the 1980 census data are available, which approach gives the more accurate prediction?

(c) Modify SPLINE so that it uses the natural end conditions, and see how this affects the prediction for 1970 through 1980. Since the natural end conditions imply the spline is a linear function outside the data interval, zero population growth cannot be predicted.

P4-6. If $P(x)$ denotes the polynomial of degree n such that $P(k) = k/(k + 1)$ for $k = 0, 1, \dots, n$, determine $P(n + 1)$.

Hints: Consider even n and odd n separately. If you cannot find an analytic solution, use the computer. (This problem, without the hints, was part of the Fourth U.S.A. Mathematical Olympiad for high school students.)

P4-7. A fictitious chemical experiment produces the following seven data points, which are known to have only negligible errors:

t	-1.000	-0.960	-0.860	-0.790	0.220	0.500	0.930
y	-1.000	-0.151	0.894	0.986	0.895	0.500	-0.306

It is desired to estimate the values of $y(t)$ for t between -1.000 and 1.000 by interpolating between the data points. It is believed that $y(t)$ is a very smooth curve.

(a) Plot the points and interpolate a smooth curve by intuition.

(b) Find and plot the unique 6th-degree polynomial which interpolates the points. See how well the polynomial fits the trend of the data.

(c) Use SPLINE to find the coefficients of the cubic spline which interpolates the data. Use SEVAL to obtain the values of the spline at enough intermediate points to plot it. See how well the spline fits the trend of the data. (We are indebted to Richard F. Thompson for suggesting this problem.)

P4-8. How many floating-point multiplications, divisions, and additions or subtractions are performed by the subroutine SPLINE? What is the total number of floating-point operations? Express your answers as functions of N.

P4-9. For the case of equally spaced knots (i.e., $h_i = h$, $i = 1, 2, \ldots, n - 1$), the algorithm used by SPLINE can be simplified. What is an algorithm for equally spaced knots? How many floating-point operations does it require? Refer to Chapter 3, Problem P3-7.

P4-10. Show that the cubic spline coefficients b_i, c_i, and d_i are given by the last three equations in Section 4.4.

P4-11. Suppose one is given n points (x_i, y_i) in the Euclidean plane ($i = 1, 2, \ldots, n$) and wants to put a smooth closed curve through them in order. One method is to draw a closed polygon joining the points in the same order and to let t stand for arc length along the polygon ($0 \leq t \leq T$), so that the vertices of the polygon occur as

$$0 < t_1 < t_2 < \ldots < t_n = T.$$

Next, one fits the data (t_i, x_i) ($i = 1, \ldots, n$) by a periodic cubic spline $x(t)$ with period T. That is, in each interval $t_i \leq t \leq t_{i+1}$, $x(t)$ is a cubic

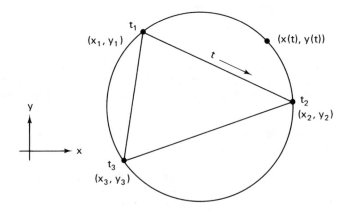

Fig. P4.11.

polynomial, and

$$x(0) = x(T), \quad x'(0) = x'(T), \quad \text{and} \quad x''(0) = x''(T).$$

Similarly, one fits the data (t_i, y_i) with a periodic cubic spline $y(t)$.

The desired curve is then given parametrically by $x(t)$ and $y(t)$.

Your task is to modify the subroutine SPLINE to compute the periodic cubic splines $x(t)$ and $y(t)$. Test the program by using as initial data the vertices of a regular polygon of n vertices for $n = 3$ and 4. See how close your solutions come to being circles.

P4-12. Let $x_i = i - 11$ for $i = 1, 2, \ldots, 21$. Let

$$y_i = \begin{cases} 1, & \text{for } i = 11, \\ 0, & \text{for } i \neq 11. \end{cases}$$

Compute, list, and graph the cubic spline through the 21 points (x_i, y_i).

Also compute, list, and graph the Lagrange polynomial $l_{11}(x)$ for the same 21 points.

Compare these important functions.

The graphs can be limited to $x \geq 0$. Why?

P4-13. Consider the function s defined as

$$s(x) = \begin{cases} 1 - 2x, & x < -3, \\ 28 + 25x + 9x^2 + x^3, & -3 \leq x < -1, \\ 26 + 19x + 3x^2 - x^3, & -1 \leq x < 0, \\ 26 + 19x + 3x^2 - 2x^3, & 0 \leq x < 3, \\ -163 + 208x - 60x^2 + 5x^3, & 3 \leq x < 4, \\ 157 - 32x, & 4 \leq x. \end{cases}$$

Show that s is a natural cubic spline function with the knots $\{-3, -1, 0, 3, 4\}$. Be sure to explicitly state each of the properties of s which are necessary for this to be true.

P4-14. Modify SEVAL so that it is a subroutine with the heading

SUBROUTINE SEVAL (N, U, X, Y, B, C, D, S, SP).

It should evaluate both the spline and its derivative at the point U and return the two values in S and SP. Test your program on some test data where the derivative is known. Also, see part (a) of Problem P7-7.

P4-15. Suppose you are given n data points, $(x_i, f(x_i))$, $i = 1, \ldots, n$, and desire to find points where $f'(x) = 0$.

(a) Why would it not be a good idea to interpolate by $p(x)$, a polynomial of degree $n - 1$, and then solve $p'(x) = 0$?

(b) Describe an efficient algorithm for solving this problem using cubic interpolatory splines.

(c) Write a program using your method and SPLINE. Try your program on the data in Problem P7-7.

5 NUMERICAL INTEGRATION

Methods of approximating functions, such as those covered in the previous chapter, provide a basis for many numerical techniques. In this chapter we shall deal primarily with the use of such approximations for calculating integrals. A few comments will also be made about the calculation of derivatives.

There are many different computer methods for integration and differentiation. The method most appropriate for a particular problem depends to a large extent on the amount of information available about the function involved. The basic problems are for a single function $f(x)$ of a single real variable x, on an interval $[a, b]$, and can be divided into four broad categories:

1. Function values $f(x_i)$ are available only for a *fixed*, finite set of points x_i in the interval $[a, b]$.

2. The function $f(x)$ is defined and can be evaluated for any *real* x in the interval $[a, b]$.

3. The definition of the function can be analytically extended to *complex x*.

4. An explicit formula for $f(x)$ is available in a form appropriate for symbolic manipulation.

Functions in the first category may result from experimental data measured at points x_i which are often unevenly spaced, or they may be obtained from tables where the x_i are evenly spaced.

For functions in the first two categories, numerical differentiation is basically more difficult than numerical integration, because numerical differentiation tends to amplify any error present in the data, but numerical integration tends to smooth and diminish such error. If the function values are known or can be computed fairly accurately, and if relatively low-order derivatives are required, then methods based on spline or polynomial interpolation are often quite satisfactory. But if high-order derivatives are required, or if the function values are *noisy*, the results may be very inaccurate.

An example of a function in the third category would be a complicated expression made up of trigonometric and other elementary functions. With problems of this type it is a good idea to take advantage of the complex extension whenever it is available. In these situations, the derivatives of f can be expressed in terms of complex contour integrals, and the smoothing effect of numerical integration can be used to produce accurate approximations to high-order derivatives. See Lyness and Moler (1967) and Lyness and Sande (1971).

For problems in the fourth category, symbolic differentiation by computer is easier than symbolic integration, just as it is in elementary calculus. See Moses (1972). Symbolic calculations of this sort should become quite common in the future; however, few people are making use of symbolic manipulation programs at the time of this writing.

The term *quadrature* is often used for numerical approximation of definite integrals in order to avoid confusion with the numerical *integration* of ordinary differential equations. The remainder of this chapter is devoted to quadrature of functions in the first two categories.

5.1. THE RECTANGLE AND TRAPEZOID RULES

Let $[a, b]$ be a finite inverval on the x axis which has been partitioned into n subintervals called *panels*, $[x_i, x_{i+1}]$, $i = 1, \ldots, n$. Assume $x_1 = a$, $x_{n+1} = b$, and $x_1 < x_2 < \ldots < x_{n+1}$. Let $h_i = x_{i+1} - x_i$ be the panel widths.

Let $f(x)$ be a function defined on $[a, b]$. Suppose an approximation is desired to the definite integral

$$I(f) = \int_a^b f(x)\, dx.$$

Clearly, $I(f)$ can be expressed as a sum of integrals over the panels

$$I(f) = \sum_{i=1}^n I_i,$$

where

$$I_i = I_i(f) = \int_{x_i}^{x_{i+1}} f(x)\, dx.$$

A *quadrature rule* is a simple formula for approximating the individual I_i. A *composite quadrature rule* is a formula giving an approximation to $I(f)$

as a sum of the quadrature rule approximations to the individual I_i. Two of the simplest quadrature rules are the rectangle rule and the trapezoid rule. In some situations, they are also among the most effective.

The rectangle rule uses function values at the midpoints of the panels

$$y_i = \frac{x_i + x_{i+1}}{2}, \quad i = 1, \ldots, n.$$

It approximates each I_i by the area of the rectangle with base h_i and height $f(y_i)$. This gives

$$I_i \approx h_i f(y_i)$$

and hence the *composite rectangle rule*:

$$R(f) = \sum_{i=1}^{n} h_i f(y_i).$$

The trapezoid rule uses function values at the end points of the panels. It approximates I_i by the area of the trapezoid with base h_i and height which varies linearly from $f(x_i)$ on the left to $f(x_{i+1})$ on the right. This leads to

$$I(f) \approx h_i \frac{f(x_i) + f(x_{i+1})}{2}$$

and hence the *composite trapezoid rule*:

$$T(f) = \sum_{i=1}^{n} h_i \frac{f(x_i) + f(x_{i+1})}{2}.$$

It is easily shown that if $f(x)$ is continuous (or even merely Riemann integrable) on $[a, b]$ and if $h = \max_i h_i$ then

$$\lim_{h \to 0} R(f) = I(f)$$

and

$$\lim_{h \to 0} T(f) = I(f).$$

That is, both rules converge as the lengths of the subintervals decrease. The crucial practical question is, How *fast* do they converge?

The rectangle rule is based on piecewise constant, or degree zero, interpolation, while the trapezoid rule is based on piecewise linear, or degree one, interpolation. One might therefore expect the trapezoid rule to be more accurate than the rectangle rule. For example, let $[a, b] = [0, 1]$, let $n = 1$ (so there is only one panel), and let $f(x) = x$. The trapezoid rule obviously gives a result with no error because the linear function interpolating $f(x)$ at any two points agrees with $f(x)$ everywhere. However, the rectangle rule also gives a result with no error despite the fact that the constant function interpolating $f(x)$ at $x = \frac{1}{2}$ does not agree with $f(x)$ at any other point. The *average* interpolation error over [0, 1] is zero.

Is this simple example typical, or is the rectangle rule just lucky? Which rule can be expected to be more accurate in general? To answer these questions, we must make a careful error analysis based on some assumptions about $f(x)$. To also use the results of the analysis in later sections, we shall include some extra terms here.

Assume that f has five continuous derivatives and that the values of these derivatives are not too large. Consider a panel $[x_i, x_{i+1}]$. The Taylor series expansion of $f(x)$ about y_i, the center of this panel, is

$$f(x) = f(y_i) + (x - y_i)f'(y_i) + \tfrac{1}{2}(x - y_i)^2 f''(y_i)$$
$$+ \tfrac{1}{6}(x - y_i)^3 f'''(y_i) + \tfrac{1}{24}(x - y_i)^4 f^{iv}(y_i) + \cdots$$

The assumptions about f imply that the remainder term denoted by "$\cdots$" is less significant than the terms explicitly shown.

To integrate this series over $[x_i, x_{i+1}]$, observe that

$$\int_{x_i}^{x_{i+1}} (x - y_i)^p \, dx = \begin{cases} h_i, & p = 0, \\ 0, & p = 1, \\ \dfrac{h_i^3}{12}, & p = 2, \\ 0, & p = 3, \\ \dfrac{h_i^5}{80}, & p = 4. \end{cases}$$

Note that the odd powers integrate to zero. Consequently,

$$\int_{x_i}^{x_{i+1}} f(x) \, dx = h_i f(y_i) + \tfrac{1}{24}h_i^3 f''(y_i) + \tfrac{1}{1920}h_i^5 f^{iv}(y_i) + \cdots$$

This shows that when h_i is small, the error for the rectangle rule on one panel is $\tfrac{1}{24}h_i^3 f''(y_i)$ plus some less significant terms.

Returning to the Taylor series and substituting $x = x_i$ and $x = x_{i+1}$, we obtain

$$f(x_i) = f(y_i) - \tfrac{1}{2}h_i f'(y_i) + \tfrac{1}{8}h_i^2 f''(y_i)$$
$$- \tfrac{1}{48}h_i^3 f'''(y_i) + \tfrac{1}{384}h_i^4 f^{iv}(y_i) + \cdots,$$

$$f(x_{i+1}) = f(y_i) + \tfrac{1}{2}h_i f'(y_i) + \tfrac{1}{8}h_i^2 f''(y_i)$$
$$+ \tfrac{1}{48}h_i^3 f'''(y_i) + \tfrac{1}{384}h_i^4 f^{iv}(y_i) + \cdots,$$

and thus

$$\frac{f(x_i) + f(x_{i+1})}{2} = f(y_i) + \tfrac{1}{8}h_i^2 f''(y_i) + \tfrac{1}{384}h_i^4 f^{iv}(y_i) + \dots.$$

Combining this with the expansion for the integral gives

$$\int_{x_i}^{x_{i+1}} f(x)\, dx = h_i \frac{f(x_i) + f(x_{i+1})}{2} - \tfrac{1}{12}h_i^3 f''(y_i) - \tfrac{1}{480}h_i^5 f^{iv}(y_i) + \dots.$$

This shows that when h_i is small, the error for the trapezoid rule on one panel is $-\tfrac{1}{12}h_i^3 f''(y_i)$ plus some less significant terms.

The total error for either rule is the sum of the errors on the individual panels. Let

$$E = \tfrac{1}{24} \sum_{i=1}^{n} h_i^3 f''(y_i),$$

$$F = \tfrac{1}{1920} \sum_{i=1}^{n} h_i^5 f^{iv}(y_i).$$

Then

$$I(f) = R(f) + E + F + \dots$$
$$= T(f) - 2E - 4F + \dots.$$

If the h_i are sufficiently small, then $h_i^5 \ll h_i^3$, so if f^{iv} is not too badly behaved, then $F \ll E$.

Several important conclusions can be drawn from these results. The first is that, for many functions $f(x)$, the rectangle rule is about twice as accurate as the trapezoid rule. We are about to develop other rules for which the error terms involve higher powers of h_i, so the factor of 2 is not very important—but it does surprise many people.

The next conclusion is that the difference between the values obtained by the rectangle and trapezoid rules can be used to estimate the error in either one of them. However, the estimate is not infallible; it is possible for both rules to give the same value but for this value to be incorrect.

Another conclusion concerns the effect of changing the number of panels. A simple change is to cut each panel in half. (We assume that function values are also given or can be computed at the new points.) Each h_i is divided by 2, and hence each h_i^3, which occurs in the dominant error term E, is decreased by a factor of $\tfrac{1}{8}$. However, the total number of panels is multiplied by 2, so the overall error term E is decreased by a factor of about $\tfrac{1}{4}$. The factor is usually not exactly $\tfrac{1}{4}$ because f'' is usually not constant and the higher-order error terms have an influence. However, in actual practice with functions having continuous, bounded second derivatives, *doubling* the number of panels in either the rectangle or trapezoid rule can be expected to roughly *quadruple* the accuracy. The difference in the results of either of these rules before and after doubling the number of panels can be used to estimate the error or improve the computed result.

The technique of repeatedly doubling the number of panels and estimating the error can be programmed to produce a method which automatically determines the panels so that the approximate integral is computed to within some prescribed error tolerance. This technique can be applied to other, more accurate, quadrature rules. We shall consider one such method in Section 5.4.

5.2. SPLINE QUADRATURE

Cubic spline interpolation can be used to obtain an interesting and useful quadrature formula. For $x_i \leq x \leq x_{i+1}$, let

$$s(x) = f_i + b_i(x - x_i) + c_i(x - x_i)^2 + d_i(x - x_i)^3$$

be the cubic spline which interpolates $f(x)$ at the nodes. Then with $a = x_1$ and $b = x_{n+1}$,

$$\int_a^b f(x)\,dx \approx \int_a^b s(x)\,dx$$

$$= \sum_{i=1}^n h_i f_i + \tfrac{1}{2}h_i^2 b_i + \tfrac{1}{3}h_i^3 c_i + \tfrac{1}{4}h_i^4 d_i.$$

The coefficients b_i, c_i, and d_i can be obtained using the subroutine SPLINE in Section 4.5. Since there are $n + 1$ nodes, SPLINE is called with $N = n + 1$.

However, it is useful to express this formula in another way. Recall from Section 4.4 that $s(x)$ can be represented on $[x_i, x_{i+1}]$ by

$$s(x) = wf_{i+1} + \bar{w}f_i + h_i^2[(w^3 - w)\sigma_{i+1} + (\bar{w}^3 - \bar{w})\sigma_i]$$

where $w = 1 - \bar{w} = (x - x_i)/h_i$ and

$$\sigma_i = \frac{s''(x_i)}{6} = \frac{c_i}{3}.$$

The σ_i are computed by solving the symmetric, tridiagonal system of equations developed in Section 4.4. To obtain the quadrature formula, observe that

$$\int_{x_i}^{x_{i+1}} s(x) \, dx = h_i \int_0^1 s(w) \, dw,$$

$$\int_0^1 w \, dw = \frac{1}{2},$$

and

$$\int_0^1 (w^3 - w) \, dw = -\frac{1}{4}.$$

Consequently,

$$\int_{x_i}^{x_{i+1}} s(x) \, dx = h_i \frac{f_i + f_{i+1}}{2} - h_i^3 \frac{\sigma_i + \sigma_{i+1}}{4}.$$

In other words, the spline quadrature formula is the same as the trapezoid rule plus a correction term involving the σ_i.

A convenient formula for actual computation is

$$\int_a^b s(x) \, dx = \sum_{i=1}^n h_i \frac{f_i + f_{i+1}}{2} - h_i^3 \frac{c_i + c_{i+1}}{12}.$$

Only the data arrays, X and F, and the second derivative array, C, computed by SPLINE are needed for the computation. However, the arrays B and D are still required within SPLINE for intermediate storage.

To assess the accuracy of spline quadrature, notice that if $f''(x)$ is not too badly behaved, then

$$h_i^3 \frac{c_i + c_{i+1}}{12} = h_i^3 \frac{s''(x_i) + s''(x_{i+1})}{24}$$

$$\approx \frac{h_i^3}{12} f''(y_i).$$

Thus the correction term provided by the spline approximates the error in the trapezoid rule itself.

We have not had any extensive practical experience using this method, but based on this analysis we expect it to be quite successful. We consequently recommend it for situations where the data points are fixed and the automatic methods described in later sections are not applicable.

It should be noted that this is *not* a conventional quadrature formula. When expressed in the form

$$\int_a^b f(x) \, dx \approx \sum_{i=1}^{n+1} \alpha_i f(x_i),$$

each of the coefficients α_i depend on *all* the function values $f(x_i)$ in a fairly complicated way.

5.3. SIMPSON'S RULE

In Section 5.1, the composite rectangle and trapezoid rules were defined by

$$R(f) = \sum_{i=1}^{n} h_i f\left(\frac{x_i + x_{i+1}}{2}\right)$$

and

$$T(f) = \sum_{i=1}^{n} h_i \frac{f(x_i) + f(x_{i+1})}{2}.$$

Furthermore, it was proved that the errors in these rules are given by

$$I(f) - R(f) = E + F + \ldots$$

and

$$I(f) - T(f) = -2E - 4F + \ldots,$$

where

$$E = \tfrac{1}{24} \sum_{i=1}^{n} h_i^3 f''(y_i),$$

$$F = \tfrac{1}{1920} \sum_{i=1}^{n} h_i^5 f^{iv}(y_i).$$

Combining these two rules in the proper way can produce a new rule for which the error formula no longer contains E. Since $R(f)$ is about twice as accurate as $T(f)$, the proper choice is

$$S(f) = \tfrac{2}{3}R(f) + \tfrac{1}{3}T(f).$$

Expanding this to display the individual terms gives

$$S(f) = \sum_{i=1}^{n} \tfrac{1}{6} h_i \left[f(x_i) + 4f\left(\frac{x_i + x_{i+1}}{2}\right) + f(x_{i+1}) \right].$$

Many readers will recognize this as the *composite Simpson's rule*. It may also be derived by integrating a piecewise parabolic function which interpolates the data.

The error in Simpson's rule can be obtained directly from our formulas for the errors in the rectangle and trapezoid rules:

$$I(f) - S(f) = \tfrac{2}{3}[I(f) - R(f)] + \tfrac{1}{3}[I(f) - T(f)]$$
$$= (\tfrac{2}{3} - \tfrac{2}{3})E + (\tfrac{2}{3} - \tfrac{4}{3})F + \dots$$
$$= -\tfrac{2}{3}F + \dots$$
$$= -\tfrac{1}{2880} \sum_{i=1}^{n} h_i^5 f^{iv}(y_i) + \dots.$$

Notice that even though $S(f)$ is based on second-degree interpolation, the error term involves the fourth derivative, and hence Simpson's rule is exact for cubic functions. In other words, like the rectangle rule, Simpson's rule obtains an "extra" order of accuracy. (The reader may verify that Simpson's rule is exact for cubics.)

If the length of each panel is halved, then each h_i^5 term in the error formula is decreased by a factor of $\tfrac{1}{32}$. The total number of terms is increased by a factor of 2, so the overall error is decreased by a factor close to $\tfrac{1}{16}$. This fact is important for programs which automatically choose the number and size of the panels.

The technique of combining two approximations with similar errors to obtain a much more accurate approximation can be continued. For example, the values of $S(f)$ for two different sets of panels can be combined to obtain a new value which differs from $I(f)$ by a quantity involving h_i^7 and $f^{(6)}(x)$. Carrying this out in a systematic way leads to a popular method known as *Romberg quadrature*.

The same idea can also be applied to numerical differentiation, to numerical solution of differential equations—in fact, to any numerical process which approximates limits and which has a known error behavior. The general technique is known as the *deferred approach to the limit* or *Richardson's extrapolation*. For an excellent survey of the development and applications of this technique, see Joyce (1971). We shall mention another example of Richardson's extrapolation in Section 6.4.

Any attempt to assess the relative merits of the methods we have described so far—namely, the rectangle, trapezoid, spline, and Simpson rules—would involve questions like "Is $h^2 f''(x)$ bigger or smaller than $h^4 f^{iv}(y)$?" The answers and resulting conclusions clearly depend on the nature of the functions being integrated. Fortunately, the computer can be used to help answer some of these questions. The resulting *adaptive quadrature routines* are discussed in the next two sections.

5.4. ADAPTIVE QUADRATURE ROUTINES

An adaptive quadrature routine is a numerical quadrature algorithm which uses one or two basic quadrature rules and which automatically determines the subinterval sizes so that the computed result meets some prescribed accuracy requirement. Different mesh sizes may be used in different parts of the interval with relatively large meshes used where the integrand is smooth and slowly varying and relatively small meshes used in regions where the integration becomes difficult. In this way, an attempt is made to provide a result with the prescribed accuracy at as small a cost in computer time as possible.

The first adaptive routine was based on Simpson's rule and was published by McKeeman (1962). This and similar routines proved very successful in practice. A fairly complete analysis of the technique and several suggestions for modifications were made by Lyness (1969a). Some connections with computer science are given by Rice (1975) and by Malcolm and Simpson (1975). In this section we shall discuss adaptive routines in general. In the next section we shall present a particular routine.

The user of one of these routines specifies a finite interval $[a, b]$, provides a subroutine which computes $f(x)$ for any x in the interval, and chooses an error tolerance ϵ. The routine attempts to compute a quantity Q, so that

$$\left| Q - \int_a^b f(x)\, dx \right| \leq \epsilon.$$

The routine may decide that the prescribed accuracy is not attainable, do the best it can, and return an estimate of the accuracy actually achieved.

In assessing the efficiency of any quadrature routine, a frequent assumption is that the major portion of the cost of the computation lies in evaluation of the integrand $f(x)$. So if two subroutines are given the same function to integrate and both produce answers of about the same accuracy, the subroutine that requires the smaller number of function evaluations is regarded as the more efficient on that particular problem.

It will always be possible to invent integrands $f(x)$ which "fool" the routine into producing a completely wrong result, but for good adaptive routines, the class of such examples has been made as small as possible without unduly complicating the logic of the routine or making it inefficient on reasonable problems.

During the computation, the interval $[a, b]$ is broken down into subintervals $[x_i, x_{i+1}]$. In most routines each subinterval is obtained by bisecting a subinterval obtained earlier in the computation. The actual number of subintervals, as well as their locations and lengths, depends on the integrand $f(x)$ and the accuracy requirement ϵ. The numbering is arranged so that $x_1 = a$ and the routine determines an n so that $x_{n+1} = b$. In this section, we shall let $h_i = x_{i+1} - x_i$ be the subinterval width.

A typical scheme applies two different quadrature rules to each subinterval. We denote the two results by P_i and Q_i. For example, schemes based on Simpson's rule use the basic two-panel formula

$$P_i = \frac{h_i}{6}\left[f(x_i) + 4f\left(x_i + \frac{h_i}{2}\right) + f(x_i + h_i)\right]$$

and a four-panel composite rule

$$Q_i = \frac{h_i}{12}\left[f(x_i) + 4f\left(x_i + \frac{h_i}{4}\right) + 2f\left(x_i + \frac{h_i}{2}\right)\right.$$
$$\left. + 4f\left(x_i + \frac{3h_i}{4}\right) + f(x_i + h_i)\right].$$

Both P_i and Q_i are approximations to

$$I_i = \int_{x_i}^{x_{i+1}} f(x)\, dx.$$

The basic idea of adaptive quadrature is to compare the two approxima-tions P_i and Q_i and thereby obtain an estimate of their accuracy. If the accu-racy is acceptable, one of them is taken as the value of the integral over the subinterval. If the accuracy is not acceptable, the subinterval is divided into two or more parts and the process repeated on the smaller subintervals.

To reduce the total number of function evaluations, it is usually arranged that the two rules producing P_i and Q_i require integrand values at some of the same points. For example, with Simpson's rule, Q_i requires five function values, three of which are also used in P_i. Consequently, processing a new subinterval requires only two new function evaluations. For the remainder of this chapter, we shall assume that Q_i is obtained simply by applying the rule for P_i twice, once to each half of the interval. This is a common tech-nique and leads to the simplest analysis.

Let us also assume that the rule for P_i gives the exact answer if the inte-grand is a polynomial of degree $p - 1$, or, equivalently, if the pth derivative $f^{(p)}(x)$ is identically zero. By expanding the integrand in a Taylor series about the midpoint of the subinterval, it then can be shown there is a constant c so that

$$I_i - P_i = ch_i^{p+1} f^{(p)}\left(x_i + \frac{h_i}{2}\right) + \cdots.$$

The exponent of h_i is $p + 1$ rather than p because the subinterval is of length h_i. From the assumption that Q_i is the sum of two P's from two subintervals of length $h_i/2$ it follows that

$$I_i - Q_i = c\left(\frac{h_i}{2}\right)^{p+1}\left[f^{(p)}\left(x_i + \frac{h_i}{4}\right) + f^{(p)}\left(x_i + \frac{3h_i}{4}\right)\right] + \cdots.$$

Since

$$f^{(p)}\left(x_i + \frac{h_i}{4}\right) + f^{(p)}\left(x_i + \frac{3h_i}{4}\right) = 2f^{(p)}\left(x_i + \frac{h_i}{2}\right) + \cdots,$$

the two errors are related by

$$I_i - Q_i = \frac{2}{2^{p+1}}(I_i - P_i) + \ldots = \frac{1}{2^p}(I_i - P_i) + \ldots.$$

This indicates that bisecting the subinterval decreases the error by a factor of about 2^p. Solving for the unknown I_i and rearranging terms, we obtain

$$Q_i - I_i = \frac{1}{2^p - 1}(P_i - Q_i) + \ldots.$$

In other words, the error in the more accurate approximation Q_i is about $1/(2^p - 1)$ times the difference between the two approximations.

The basic task of a typical routine is to bisect each subinterval until the following inequality is satisfied:

$$\frac{1}{2^p - 1}|P_i - Q_i| \le \frac{h_i}{b - a}\epsilon,$$

where ϵ is the user-supplied accuracy tolerance. If the entire interval $[a, b]$ can be covered by n subintervals on which this is valid, then the routine would return

$$Q = \sum_{i=1}^{n} Q_i.$$

Combining the above and ignoring the higher-order terms, we find

$$\left| Q - \int_a^b f(x)\,dx \right| = \left| \sum_{i=1}^{n} Q_i - I_i \right|$$

$$\le \sum_{i=1}^{n} |Q_i - I_i|$$

$$\le \frac{1}{2^p - 1} \sum_{i=1}^{n} |P_i - Q_i|$$

$$\le \frac{1}{2^p - 1} \cdot \frac{2^p - 1}{b - a} \epsilon \sum_{i=1}^{n} h_i$$

$$= \epsilon,$$

which is the desired goal.

This analysis requires the assumptions that $f^{(p)}(x)$ is continuous and that the error is proportional to $h_i^{p+1} f^{(p)}(x)$. Even when this is not exactly satisfied, the routine may nevertheless perform well and the final result may be within the desired tolerance, although the detailed behavior will be different.

For simplicity, we have also assumed that the routine is using an absolute error criterion, that is,

$$\left| Q - \int f \right| \le \epsilon.$$

In many situations, some kind of relative error tolerance which is independent of scale factors in f is preferable. A pure relative criterion,

$$\frac{\left| Q - \int f \right|}{\left| \int f \right|} \le \epsilon,$$

is complicated by several factors. The denominator, $\int f$, might be zero. Even if the denominator is merely close to zero because positive values of f on one part of the interval nearly cancel negative values of f on another part, the criterion may be impossible to satisfy in practice. Furthermore, a good approximation to the value of the denominator is not available until the end of the computation.

Some routines use a criterion involving $\int |f|$,

$$\frac{\left| Q - \int f \right|}{\int |f|} \le \epsilon.$$

The denominator cannot be zero unless f is identically zero and does not suffer from the cancellation difficulties associated with oscillatory integrands. However, it does involve more computation, although no more function evaluations, and does require a more elaborate explanation.

The users of adaptive quadrature routines should be aware of these various accuracy tests and check the documentation of a particular routine to see which one is used.

The behavior of a typical adaptive routine on an interesting example is shown in Fig. 5.1. The problem is to integrate

$$f(x) = \frac{1}{(x - 0.3)^2 + 0.01} + \frac{1}{(x - 0.9)^2 + 0.04} - 6$$

over the interval [0, 1]. QUANC8 was used on an IBM 360 with long arithmetic and RELERR $= 10^{-15}$ (see the next section). Statements were added to the routine to call a plot subroutine with each of the points $(x_i, f(x_i))$ and $(x_i, 0)$, $i = 1, \ldots, n$. The integrand has a high peak near $x = 0.3$ and a smaller peak near $x = 0.9$. A graph of the relevant derivative $f^{(10)}(x)$ would also have maxima at these points. The figure shows that the subroutine has

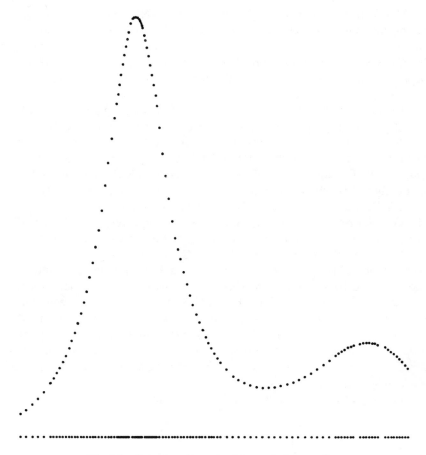

Fig. 5.1. Behavior of an adaptive quadrature routine.

chosen very small intervals in the vicinity of the peaks and larger intervals elsewhere.

This behavior is fairly typical of QUANC8, as well as of other adaptive routines. If the interval spacing which is required near the peaks to obtain the desired accuracy were used throughout the entire interval, considerably more computing time would be required.

5.5. SUBROUTINE QUANC8

Over a dozen reasonably effective adaptive quadrature routines have been published as of 1976. Two of the most widely used are SQUANK, written by James Lyness, and QNC7, written by David Kahaner and Rondall Jones. In this section, we shall present yet another adaptive routine,

QUANC8. It combines the basic ideas in SQUANK and QNC7 but does not include some of their more sophisticated features, such as control of roundoff error and detection of certain types of singularities. We believe that QUANC8 is fairly easy to read and understand and that its performance on a wide range of problems is still competitive with more complicated routines.

The user of QUANC8 must supply a function subprogram FUN(X) for the integrand $f(x)$; the lower and upper limits of integration, $A = a$ and $B = b$; and the absolute and relative error tolerances, ABSERR and RELERR. The routine returns RESULT = the approximation to the integral, ERREST = an estimate of the absolute error in RESULT, NOFUN = the number of function evaluations required, and FLAG, a reliability indicator. If FLAG is zero, then RESULT probably has the desired accuracy. If FLAG is nonzero but small, RESULT may still be acceptable. If FLAG is large, then some kind of unacceptable integrand has been encountered.

The name QUANC8 is derived from Quadrature, Adaptive, Newton-Cotes' 8-panel. The *Newton-Cotes formulas* are the family of quadrature rules obtained by integrating interpolating polynomials over equally spaced evaluation points. The rectangle, trapezoid, and Simpson's rules are obtained by integrating zeroth-, first-, and second-degree polynomials and are the first three members of the family. The 8-panel formula is obtained by integrating an eighth-degree polynomial, but just as with the rectangle and Simpson's rules, an "extra" degree is achieved, so the formula gives the exact result for polynomials of degree 9.

For integration over the interval [0, 1] the rule is

$$\int_0^1 f(x)\, dx \approx \sum_{k=0}^8 w_k f\left(\frac{k}{8}\right).$$

Approximate values of the weights w_k are

k	w_k
0	0.0349
1	0.2077
2	−0.0327
3	0.3702
4	−0.1602
5	0.3702
6	−0.0327
7	0.2077
8	0.0349

Rational expressions for the actual weights (multiplied by 8) are included in the subroutine itself. Since the rule is exact for $f(x) \equiv 1$, the weights satisfy

$$\sum_{k=0}^{8} w_k = 1.$$

However, some of the weights are negative, and

$$\sum_{k=0}^{8} |w_k| = 1.4512.$$

This factor is important in error analysis. For example, suppose the function $f(x) \equiv 1$ were contaminated with some sort of error so that $f(x) = 0.99$ at $x = \frac{2}{8}, \frac{4}{8}, \frac{6}{8}$, while $f(x) = 1.01$ at the other nodes. Then the rule would give 1.014512 as the value of the integral. Errors in the function values are slightly magnified in the result. The magnification factor is inconsequential in this rule, but for Newton-Cotes' formulas based on polynomials of higher degree, it becomes unacceptable.

We have chosen a rule with eight panels so that each panel width is the basic interval width, $b - a$, divided by a power of 2. Consequently, on a binary computer there are usually no roundoff errors in computing the nodes x_i.

The 8-panel rule is used to obtain the quantity P_i of the previous section. The quantity Q_i is obtained by applying the same rule to the two halves of the interval, thereby using 16 panels. In the program, the variables QPREV and QNOW are used for P_i and Q_i. The variables QLEFT and QRIGHT are used for the results on the two halves of the interval.

Since the basic rule is exact for ninth-degree polynomials, the analysis of the previous section applies with $p = 10$. The error in Q_i will be about $1/(2^{10} - 1) = \frac{1}{1023}$ times the error in P_i whenever the integrand is smooth and higher-order terms can be ignored.

As was indicated in the last section, the results for a particular subinterval are acceptable if

$$\frac{1}{1023}|P_i - Q_i| \leq \frac{h_i}{b-a}\epsilon.$$

In QUANC8, this test is made by comparing two quantities

```
ESTERR = ABS(QNOW − QPREV)/1023
TOLERR = AMAX1(ABSERR, RELERR*ABS(AREA))*(STEP/STONE).
```

The ratio STEP/STONE is equal to $h_i/(b - a)$, and AREA is an estimate of the integral over the entire interval, so ϵ is the maximum of ABSERR and RELERR*$|\int f|$. The subinterval is acceptable if

ESTERR .LE. TOLERR.

When the subinterval is accepted, QNOW is added to the output parameter RESULT, and ESTERR is added to the output parameter ERREST. Furthermore, the quantity (QNOW − QPREV)/1023 is added to the internal variable COR11, because $Q_i + (Q_i − P_i)/1023$ is usually a more accurate estimate of I_i than simply Q_i. In fact, this is one step of Richardson extrapolation. While the individual rules P_i and Q_i are exact for polynomials of degree 9, the combination of the two rules gives the exact value for integrals of polynomials of degree 11. The final value of COR11 is added to RESULT just before the return.

Each time a subinterval is bisected, the nodes and function values for the right half are saved for later use. Since a maximum amount of storage for these values must be specified, a limit LEVMAX = 30 is set on the level of bisection. When this maximum is reached, the subinterval is accepted even if the estimated error is too large, but a count of the number of such subintervals is kept and returned as the integer part of FLAG. The length of each of these subintervals is quite small, $(b − a)/2^{30}$, so, for most integrands, a few of them can be included in the final result without affecting its accuracy too badly.

However, if the integrand has discontinuities or unbounded derivatives or is contaminated by roundoff error, the bisection level maximum will be reached quite frequently. Consequently, another limit, NOMAX, is set on the number of function evaluations. When it appears that this limit may be exceeded, LEVMAX is reduced to LEVOUT so that the remainder of the interval can be processed with fewer than NOMAX function evaluations. Furthermore, the point X0 where the trouble is encountered is noted and the quantity

$$(B − X0)/(B − A)$$

is returned as the fractional part of FLAG. This is the portion of the interval that remains to be processed with the lower bisection limit. Since

$$X0 = B − (\text{fractional part of FLAG}) * (B − A)$$

the location of the difficulty can be determined.

QUANC8 is based on piecewise polynomial approximation and hence is not designed to handle certain kinds of integrals. Roughly, these are integrals of functions $f(x)$ for which some derivative $f^{(k)}(x)$ with $k \leq 10$ is unbounded or fails to exist. For example, integrals of the form

$$\int_0^1 x^\alpha \, dx$$

with $\alpha > −1$ are well defined, but if α is not an integer, QUANC8 may use a large number of function evaluations and may not produce an accurate error estimate.

Either of the error tolerances may be set to zero, but if they are both zero, the maximum function count will be reached unless the integrand is very simple.

The following sample program computes Si(2), where Si(z) is the sine integral,

$$Si(z) = \int_0^z \frac{\sin(x)}{x}\, dx.$$

The declaration in the main program that FUN is EXTERNAL is essential. Since the main program is compiled separately from the subprograms, there is no other indication that FUN is not a simple real variable. This is a fairly easy problem, and QUANC8 requires only 33 function evaluations. The output produced on the IBM 360 using long arithmetic is

RESULT = 1.6054129768   ERREST = 0.23D-15

If the problem is changed to

$$\int_0^2 \frac{\tan(x)}{x}\, dx,$$

the behavior is quite different. The integrand has a nonintegrable singularity at $x = \pi/2 \approx 1.57$. QUANC8 returns with FLAG = 91.21. This indicates that 91 subintervals did not converge and that 0.21 of the interval remained to be processed when trouble was detected. Since $2 - 0.21*2 = 1.58$, the singularity is readily found.

```
C  SAMPLE PROGRAM FOR QUANC8
C
      REAL FUNCTION FUN (X)
      REAL X
      IF (X .EQ. 0.0) FUN = 1.0
      IF (X .NE. 0.0) FUN = SIN(X)/X
      RETURN
      END
C
      EXTERNAL FUN
      REAL A,B,ABSERR,RELERR,RESULT,ERREST,FLAG
      INTEGER  NOFUN
      A = 0.0
      B = 2.0
      RELERR = 1.0E-10
      ABSERR = 0.0
      CALL QUANC8(FUN,A,B,ABSERR,RELERR,RESULT,ERREST,NOFUN,FLAG)
      WRITE(6,1) RESULT, ERREST
      IF (FLAG .NE. 0.0) WRITE(6,2) FLAG
    1 FORMAT(8H RESULT=, F15.10, 10H    ERREST=, E10.2)
    2 FORMAT(44H WARNING..RESULT MAY BE UNRELIABLE.  FLAG = ,F6.2)
      STOP
      END
```

```
      SUBROUTINE QUANC8(FUN,A,B,ABSERR,RELERR,RESULT,ERREST,NOFUN,FLAG)
C
      REAL FUN, A, B, ABSERR, RELERR, RESULT, ERREST, FLAG
      INTEGER NOFUN
C
C     ESTIMATE THE INTEGRAL OF FUN(X) FROM A TO B
C     TO A USER PROVIDED TOLERANCE.
C     AN AUTOMATIC ADAPTIVE ROUTINE BASED ON
C     THE 8-PANEL NEWTON-COTES RULE.
C
C     INPUT ..
C
C     FUN     THE NAME OF THE INTEGRAND FUNCTION SUBPROGRAM FUN(X).
C     A       THE LOWER LIMIT OF INTEGRATION.
C     B       THE UPPER LIMIT OF INTEGRATION.(B MAY BE LESS THAN A.)
C     RELERR  A RELATIVE ERROR TOLERANCE. (SHOULD BE NON-NEGATIVE)
C     ABSERR  AN ABSOLUTE ERROR TOLERANCE. (SHOULD BE NON-NEGATIVE)
C
C     OUTPUT ..
C
C     RESULT  AN APPROXIMATION TO THE INTEGRAL HOPEFULLY SATISFYING THE
C             LEAST STRINGENT OF THE TWO ERROR TOLERANCES.
C     ERREST  AN ESTIMATE OF THE MAGNITUDE OF THE ACTUAL ERROR.
C     NOFUN   THE NUMBER OF FUNCTION VALUES USED IN CALCULATION OF RESULT.
C     FLAG    A RELIABILITY INDICATOR.  IF FLAG IS ZERO, THEN RESULT
C             PROBABLY SATISFIES THE ERROR TOLERANCE.  IF FLAG IS
C             XXX.YYY , THEN  XXX = THE NUMBER OF INTERVALS WHICH HAVE
C             NOT CONVERGED AND  0.YYY = THE FRACTION OF THE INTERVAL
C             LEFT TO DO WHEN THE LIMIT ON  NOFUN  WAS APPROACHED.
C
      REAL W0,W1,W2,W3,W4,AREA,X0,F0,STONE,STEP,COR11,TEMP
      REAL QPREV,QNOW,QDIFF,QLEFT,ESTERR,TOLERR
      REAL QRIGHT(31),F(16),X(16),FSAVE(8,30),XSAVE(8,30)
      INTEGER LEVMIN,LEVMAX,LEVOUT,NOMAX,NOFIN,LEV,NIM,I,J
C
C     ***   STAGE 1 ***    GENERAL INITIALIZATION
C     SET CONSTANTS.
C
      LEVMIN = 1
      LEVMAX = 30
      LEVOUT = 6
      NOMAX = 5000
      NOFIN = NOMAX - 8*(LEVMAX-LEVOUT+2**(LEVOUT+1))
C
C     TROUBLE WHEN NOFUN REACHES NOFIN
C
      W0 =    3956.0 / 14175.0
      W1 =   23552.0 / 14175.0
      W2 =   -3712.0 / 14175.0
      W3 =   41984.0 / 14175.0
      W4 =  -18160.0 / 14175.0
```

```
C
C    INITIALIZE RUNNING SUMS TO ZERO.
C
      FLAG   = 0.0
      RESULT = 0.0
      COR11  = 0.0
      ERREST = 0.0
      AREA   = 0.0
      NOFUN  = 0
      IF (A .EQ. B) RETURN
C
C    ***    STAGE 2 ***    INITIALIZATION FOR FIRST INTERVAL
C
      LEV = 0
      NIM = 1
      X0 = A
      X(16) = B
      QPREV  = 0.0
      F0 = FUN(X0)
      STONE = (B - A) / 16.0
      X(8)  =  (X0  + X(16)) / 2.0
      X(4)  =  (X0  + X(8))  / 2.0
      X(12) =  (X(8)  + X(16)) / 2.0
      X(2)  =  (X0 + X(4))   / 2.0
      X(6)  =  (X(4)  + X(8))  / 2.0
      X(10) =  (X(8)  + X(12)) / 2.0
      X(14) =  (X(12) + X(16)) / 2.0
      DO 25 J = 2, 16, 2
         F(J) = FUN(X(J))
   25 CONTINUE
      NOFUN = 9
C
C    ***    STAGE 3 ***    CENTRAL CALCULATION
C    REQUIRES QPREV,X0,X2,X4,...,X16,F0,F2,F4,...,F16.
C    CALCULATES X1,X3,...X15, F1,F3,...F15,QLEFT,QRIGHT,QNOW,QDIFF,AREA.
C
   30 X(1) = (X0 + X(2)) / 2.0
      F(1) = FUN(X(1))
      DO 35 J = 3, 15, 2
         X(J) = (X(J-1) + X(J+1)) / 2.0
         F(J) = FUN(X(J))
   35 CONTINUE
      NOFUN = NOFUN + 8
      STEP = (X(16) - X0) / 16.0
      QLEFT   =  (W0*(F0 + F(8))  + W1*(F(1)+F(7))  + W2*(F(2)+F(6))
     1     + W3*(F(3)+F(5))   +   W4*F(4)) * STEP
      QRIGHT(LEV+1)=(W0*(F(8)+F(16))+W1*(F(9)+F(15))+W2*(F(10)+F(14))
     1     + W3*(F(11)+F(13)) + W4*F(12)) * STEP
      QNOW = QLEFT + QRIGHT(LEV+1)
      QDIFF = QNOW - QPREV
      AREA = AREA + QDIFF
```

```
C
C   ***   STAGE 4 *** INTERVAL CONVERGENCE TEST
C
      ESTERR = ABS(QDIFF) / 1023.0
      TOLERR = AMAX1(ABSERR,RELERR*ABS(AREA)) * (STEP/STONE)
      IF (LEV .LT. LEVMIN) GO TO 50
      IF (LEV .GE. LEVMAX) GO TO 62
      IF (NOFUN .GT. NOFIN) GO TO 60
      IF (ESTERR .LE. TOLERR) GO TO 70
C
C   ***   STAGE 5   ***   NO CONVERGENCE
C   LOCATE NEXT INTERVAL.
C
   50 NIM = 2*NIM
      LEV = LEV+1
C
C   STORE RIGHT HAND ELEMENTS FOR FUTURE USE.
C
      DO 52 I = 1, 8
         FSAVE(I,LEV) = F(I+8)
         XSAVE(I,LEV) = X(I+8)
   52 CONTINUE
C
C   ASSEMBLE LEFT HAND ELEMENTS FOR IMMEDIATE USE.
C
      QPREV = QLEFT
      DO 55 I = 1, 8
         J = -I
         F(2*J+18) = F(J+9)
         X(2*J+18) = X(J+9)
   55 CONTINUE
      GO TO 30
C
C   ***   STAGE 6   ***   TROUBLE SECTION
C   NUMBER OF FUNCTION VALUES IS ABOUT TO EXCEED LIMIT.
C
   60 NOFIN = 2*NOFIN
      LEVMAX = LEVOUT
      FLAG = FLAG + (B - X0) / (B - A)
      GO TO 70
C
C   CURRENT LEVEL IS LEVMAX.
C
   62 FLAG = FLAG + 1.0
C
C   ***   STAGE 7   ***   INTERVAL CONVERGED
C   ADD CONTRIBUTIONS INTO RUNNING SUMS.
C
   70 RESULT = RESULT + QNOW
      ERREST = ERREST + ESTERR
      COR11  = COR11  + QDIFF / 1023.0
```

```
C
C   LOCATE NEXT INTERVAL.
C
   72 IF (NIM .EQ. 2*(NIM/2)) GO TO 75
      NIM = NIM/2
      LEV = LEV-1
      GO TO 72
   75 NIM = NIM + 1
      IF (LEV .LE. 0) GO TO 80
C
C   ASSEMBLE ELEMENTS REQUIRED FOR THE NEXT INTERVAL.
C
      QPREV = QRIGHT(LEV)
      X0 = X(16)
      F0 = F(16)
      DO 78 I = 1, 8
         F(2*I) = FSAVE(I,LEV)
         X(2*I) = XSAVE(I,LEV)
   78 CONTINUE
      GO TO 30
C
C   ***   STAGE 8   ***   FINALIZE AND RETURN
C
   80 RESULT = RESULT + COR11
C
C   MAKE SURE ERREST NOT LESS THAN ROUNDOFF LEVEL.
C
      IF (ERREST .EQ. 0.0) RETURN
   82 TEMP = ABS(RESULT) + ERREST
      IF (TEMP .NE. ABS(RESULT)) RETURN
      ERREST = 2.0*ERREST
      GO TO 82
      END
```

PROBLEMS

P5-1. The integral defining the error function

$$\text{erf}(x) = \frac{2}{\sqrt{\pi}} \int_0^x e^{-t^2}\, dt$$

is fairly easy to evaluate numerically. Write a program which uses QUANC8 to print a table of erf(x) for $x = 0.0, 0.1, 0.2, \ldots, 1.9, 2.0$. Compare your table with published values or with values available from a reliable subroutine on your computer.

Since the calculation of each entry in your table is done independently of the other entries, this is not a particularly efficient way to generate such a table. For a better way, see Problem P6-1.

P5-2. Compute approximations to π by using numerical integration to evaluate

$$\pi = \int_0^1 \frac{4}{1 + x^2}\, dx.$$

(a) Use the trapezoid rule and the rectangle rule with panels of uniform length $h = 1/n$. Try $n = 8, 32, 128$. Observe that the error is approximately proportional to h^2.

(b) Use spline quadrature with the same values of h. Is the error approximately proportional to some power of h?

(c) Use QUANC8 with various values of the error tolerances. Print out NOFUN, and compare the number of points needed to obtain a certain accuracy with those in parts (a) and (b). Since this is a fairly easy problem, the error tolerances should not affect the number of points.

P5-3. Use the definition of the Riemann integral and the intermediate value theorem to prove that the composite rectangle and trapezoid rules converge to the integral as $h \longrightarrow 0$. State explicitly what assumptions you are making about the integrand.

P5-4. Suppose QUANC8 were used to integrate a function which is identically zero, say over the interval [0, 1]. The result, of course, will be zero. At what points will QUANC8 call for evaluation of the integrand? Try to answer this question by simply reading QUANC8. Then verify your answer by running a program with the following function subprogram:

```
        REAL FUNCTION FUN(X)
        REAL X
        WRITE(6,1) X
    1   FORMAT(F12.8)
        FUN = 0.0
        RETURN
        END
```

P5-5. (a) What result will be produced by using QUANC8 for

$$\int_0^{2\pi} [1 - \cos(32x)] \, dx?$$

What is the correct answer? How could QUANC8 be used to get a better answer? Again, try to answer these questions without actually running a program.

(b) Assume you are given some other quadrature routine. Explain how to find a smooth integrand with no singularities which will "fool" the routine.

P5-6. Modify QUANC8 so that it returns or prints out the total number of function evaluations and the length of the smallest panel which it uses. Run it on a fairly hard problem, such as $f(x) = \sqrt{x}$ or $f(x) = 1/((x - c)^2 + \epsilon)$ with c in the interval of integration and ϵ very small. Compare the number of evaluations actually required with the number that would be required if the smallest panel were used throughout the interval.

P5-7. Find the weights in the 8-panel Newton-Cotes' formula by requiring that

$$\int_0^1 f(x) \, dx \approx \sum_{k=0}^{8} w_k f\left(\frac{k}{8}\right)$$

is exact for $f(x) = x^p, p = 0, 1, \ldots, 8$. Use DECOMP and SOLVE. Show that the formula is also exact for $f(x) = x^9$. Show that the use of COR11 in QUANC8 makes RESULT exact if $f(x) = x^{10}$ and x^{11}.

P5-8. Write a subroutine SPLINT whose input is N data points X(I), Y(I), $I = 1, \ldots, N$, and whose output is the spline approximation to the integral

$$\int_{X(1)}^{X(N)} y(x) \, dx$$

described in Section 5.2. Your subroutine should call SPLINE.

P5-9. Which of the following problems would tend to be hard, in the sense of requiring many function evaluations, for QUANC8 with RELERR = 10^{-8}? Why? Try to answer without actually running them.

(a) $\displaystyle\int_0^1 e^{x^2} \, dx.$

(b) $\displaystyle\int_0^2 \sin(10x) \, dx.$

(c) $\displaystyle\int_0^1 \frac{\sin x}{x - 3} \, dx.$

(d) $\displaystyle\int_1^5 \frac{(x - 1)^{1/5}}{x^2 + 1} \, dx.$

(e) $\displaystyle\int_1^{10} \ln x \, dx.$

(f) $\int_0^4 f(x)\, dx$, where $f(x) = \begin{cases} 1, & x = 0, \\ \dfrac{(e^x - 1)^5}{x^5}, & x \neq 0. \end{cases}$

P5-10. Describe an efficient and accurate method to estimate $\int_0^4 f(x)\, dx$, where

$$f(x) = \begin{cases} e^{x^2}, & 0 \le x \le 2. \\ \dfrac{1}{4 - \sin 16\pi x}, & 2 < x \le 4. \end{cases}$$

P5-11. The performance of an adaptive quadrature routine can be somewhat improved using a technique developed by David Kahaner called *banking*. Banking exploits the fact that when a subinterval is accepted its error estimate is usually somewhat less than the tolerance. The difference between the tolerance and the estimate can be "deposited into a bank" and used later for increasing the effective error tolerance of other subintervals. The criteria for subinterval acceptance becomes

$$\text{ESTERR .LE. TOLERR + THETA*BANK.}$$

The value of THETA is chosen between 0 and 1 to restrict the maximum "withdrawal" from the BANK.

Modify QUANC8 to include banking. Experiment with different values of THETA on various integrands.

Using ABSERR = 0. and RELERR = 1.D–15, run a double-precision version of your modified QUANC8 subroutine on the integrals

$$\int_0^1 \frac{dx}{(x - 0.1)^2 + 0.0001}$$

and

$$\int_1^0 \frac{dx}{(x - 0.1)^2 + 0.0001}.$$

Compute each integral for THETA values of 0., .1, .2, ..., 1. Note that the case THETA = 0. is the same subinterval acceptance criterion as the unmodified QUANC8. Why does the banking help more with the second integral than with the first?

Values of THETA greater than 1. can be used to "extend credit" which must be "paid back" to the bank by later subintervals. Try extending credit to each of the above problems.

P5-12. A common problem of applied mathematics is that of solving the integral equation

$$f(x) + \int_a^b K(x, t)y(t)\, dt = y(x),$$

where the functions $f(x)$ and $K(x, t)$ are given and the problem is to compute $y(x)$.

If we approximate the integral by the quadrature formula

$$\int_a^b K(x, t)y(t)\, dt \approx \sum_{i=1}^n \alpha_i K(x, t_i)y(t_i),$$

then the integral equation becomes a system of linear algebraic equations:

$$f(t_j) + \sum_{i=1}^n \alpha_i K(t_j, t_i)y(t_i) = y(t_j), \qquad j = 1, 2, \ldots, n.$$

The solution $y(t_i)$, $i = 1, \ldots, n$, is the desired discretized approximation to the function $y(t)$.

Using Simpson's rule, find an approximate solution of the integral equation

$$\frac{4x^3 + 5x^2 - 2x + 5}{8(x + 1)^2} + \int_0^1 \left(\frac{1}{1 + t} - x\right)y(t)\, dt = y(x).$$

Fit a spline function to the discretized approximation of the function $y(t)$ using the subroutine SPLINE in Chapter 4. Compare the resulting spline function with the true solution

$$y(x) = (1 + x)^{-2}$$

at various (tabular and nontabular) points in the interval [0, 1].

6 INITIAL VALUE PROBLEMS IN ORDINARY DIFFERENTIAL EQUATIONS

6.1. THE PROBLEM TO BE SOLVED

A first-order differential equation may be written as

$$y' = f(y, t).$$

This equation has a family of solution curves, $y(t)$. For example, if $f(y, t) = y$, then for any constant C, the function $y(t) = Ce^t$ is a solution. The choice of an initial value, say $y(0)$, serves to select one of the curves of the family, as shown in Fig. 6.1. The initial value of the dependent variable could be specified for any value of the independent variable, say t_0. However, it is often assumed that a transformation has been made so that $t_0 = 0$. This does not affect the solution or the methods used to approximate the solution.

Often more than one dependent variable is involved and the problem is to solve a system of first-order equations; for example,

$$y' = f(y, z, t),$$
$$z' = g(y, z, t).$$

Assume that $\partial f/\partial y$, $\partial f/\partial z$, $\partial g/\partial y$, and $\partial g/\partial z$ exist throughout the interval of integration. The solution of this system involves two constants of integration, and so two additional pieces of information must be given to determine these constants of integration. If the values of y and z are specified at one value of the independent variable, t_0, then there will be a unique solution to the system. The problem of determining values of y and z for (future) values of $t > t_0$ is called an *initial value problem*.

Any nth-order ordinary differential equation that can be written with the highest-order derivative as the left-hand side and appearing nowhere else may

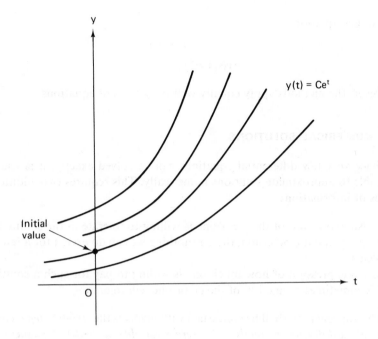

Fig. 6.1. Family of solutions of $y' = y$.

also be written as a system of n first-order equations by defining $n - 1$ new variables. For example,

$$u'' = g(u, u', t)$$

can be written as the system

$$z' = g(u, z, t),$$
$$u' = z,$$

where $z'(t) = u''(t)$. This may be written in vector notation as

$$y' = f(y, t),$$

where

$$y = \begin{pmatrix} z \\ u \end{pmatrix} = \begin{pmatrix} y_1 \\ y_2 \end{pmatrix},$$
$$f(y, t) = \begin{pmatrix} g(y_2, y_1, t) \\ y_1 \end{pmatrix}.$$

In discussing methods for the initial value problem, it is convenient to think of a single equation

$$y' = f(y, t),$$
$$y(t_0) = y_0.$$

However, the methods apply equally well to systems of equations.

6.2. NUMERICAL SOLUTIONS

Since very few differential equations can be solved exactly, it is usually desirable to approximate solutions numerically. This requires two additional pieces of information:

1. An expression of the error one is willing to tolerate in the solution. If one requires infinite precision, there is nothing we can do except for a few toy problems.
2. An expression of how much one is willing to pay for such a solution. This is sometimes a function of the error one will tolerate.

The approach we shall be concerned with involves the *step-by-step methods* (also called *difference methods* and *discrete variable methods*). A sequence of discrete points $t_0, t_1, t_2, \ldots$ is generated, possibly with variable spacing, $h_n = t_{n+1} - t_n$. At each point t_n, the solution $y(t_n)$ is approximated by a number y_n which is computed from earlier values. A difference method which provides a rule for computing y_{n+1} using k earlier values $y_n, y_{n-1}, \ldots, y_{n-k+1}$ is called a *k-step method*. If $k = 1$, it is a *single-step* method, and if $k > 1$, it is a *multistep* method.

An example of a single-step method is *Euler's* method. In Euler's method the value of y_{n+1} is calculated by a straight-line extrapolation from the previous point y_n. Consider the single equation

$$y' = f(y, t),$$

and suppose that $y(t_0) = y_0$ is given. The slope of the solution $y(t)$ can be calculated at the initial condition from $y_0' = f(y_0, t_0)$. Then an approximation y_1 to $y(t_1)$ can be calculated using the first two terms of Taylor's series:

$$y(t_1) \approx y_1 = y_0 + h_0 f(y_0, t_0).$$

We then let $t_2 = t_1 + h_1$ and calculate

$$y(t_2) \approx y_2 = y_1 + h_1 f(y_1, t_1),$$

and so on. In general,

$$y_{n+1} = y_n + h_n f(y_n, t_n),$$

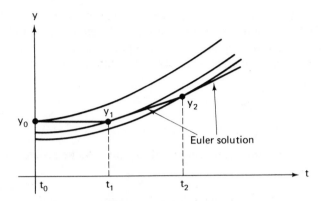

Fig. 6.2. Euler's method.

At each step the approximate solution usually crosses over onto another member of the family of solutions, as shown in Fig. 6.2. This phenomenon can lead to large errors with some differential equations. For example, the early errors made with the Euler method in solving $y' = y$ (see Fig. 6.1) are magnified by a factor e^t as time increases. We call this phenomenon *instability of the differential equation*. We can sometimes get around this difficulty by solving the problem backward in time. However, when a system of equations must be solved, the system is often unstable regardless of the direction of solution.

If, on the other hand, we have the equation $y' = -y$, we get the family of curves shown in Fig. 6.3. In this case the initial errors decrease as t increases. This is called *stability of the differential equation*. Thus, for the differential equation $y' = \lambda y$, where λ is given, initial errors in the solution are magnified

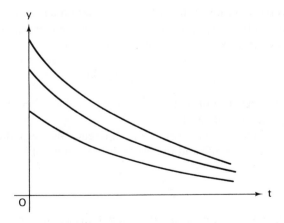

Fig. 6.3. Family of solutions of $y' = -y$.

by $e^{\lambda t}$ as t increases. If $\lambda \leq 0$, the initial errors are not increased, so the equation is stable. If $\lambda > 0$, the equation is unstable. A stable differential equation is not necessarily an easy equation to solve numerically, as we shall later see. If we want a low *relative* error, stability of the differential equation may not matter.

6.3. ERRORS

Errors enter the numerical solution of the initial value problem from two sources:

1. *Discretization error* (sometimes called *truncation* error).
2. *Roundoff error*.

Discretization error is a property of the method used. That is, if all arithmetic calculations could be performed with infinite precision, discretization error would be the only error present. Roundoff error is a property of the computer and the program. Since the exact solution is generally not known and cannot be calculated, the discretization error is usually estimated or bounded.

It is important to distinguish between two measures of discretization error, the *local* discretization error and the *global* discretization error. The local discretization error is the error that would be made in one step if the previous values were exact and if there were no roundoff error. More precisely, let $u_n(t)$ be a function of t defined by

$$u'_n = f(u_n, t),$$
$$u_n(t_n) = y_n.$$

That is, $u_n(t)$ is the solution of the differential equation determined not by the original initial condition at t_0 but by the value of the computed solution at t_n. The local discretization error d_n is then defined to be

$$d_n = y_{n+1} - u_n(t_{n+1}).$$

It is the difference between the computed solution and the theoretical solution determined by the same data at t_n.

The global discretization error is the difference between the computed solution (ignoring roundoff) and the true solution determined by the original data at t_0, that is,

$$e_n = y_n - y(t_n).$$

The distinction between local and global discretization error can be easily seen in the special case where $f(y, t)$ does not depend on y. In this case, the solution is simply an integral, $y(t) = \int_{t_0}^{t} f(\tau)\,d\tau$. Euler's method becomes a scheme for numerical quadrature that might be called "the lazy man's rectangle rule," which uses function values at the ends of the subintervals rather than at the midpoints:

$$\int_{t_0}^{t_N} f(\tau)\,d\tau \approx \sum_{n=0}^{N-1} h_n f(t_n).$$

The local discretization error is the error in one subinterval

$$d_n = h_n f(t_n) - \int_{t_n}^{t_{n+1}} f(\tau)\,d\tau,$$

and the global discretization error is the total error

$$e_N = \sum_{n=0}^{N-1} h_n f(t_n) - \int_{t_0}^{t_N} f(\tau)\,d\tau.$$

In this special case, each of the subintegrals is independent of the others (the sum could be evaluated in any order), and so the global error is the sum of the local errors:

$$e_N = \sum_{n=0}^{N-1} d_n.$$

In the case of a genuine differential equation where $f(y, t)$ depends on y, the error in any one interval depends on the solutions computed for earlier intervals. Consequently, the global error will generally be larger than the sum of the local errors if the differential equation is unstable, but it will be less than the sum if the differential equation is stable. Careful study of Figs. 6.4 and 6.5 should clarify these concepts.

The stability of a differential equation, and hence the effect of local error on global error, is governed by the sign of $\partial f/\partial y$. For nonlinear equations, $\partial f/\partial y$ may change sign, and so the equation may be unstable in some regions and stable in others. For systems of nonlinear equations, the situation is even more complex.

A fundamental concept in assessing the accuracy of a numerical method is its *order*. The order is defined in terms of the local discretization error obtained when the method is applied to problems with smooth solutions. A method is said to be of order p if there is a number C so that

$$|d_n| \leq Ch_n^{p+1}.$$

The number C may depend on the derivatives of the function defining the differential equation and on the length of the interval over which the solution

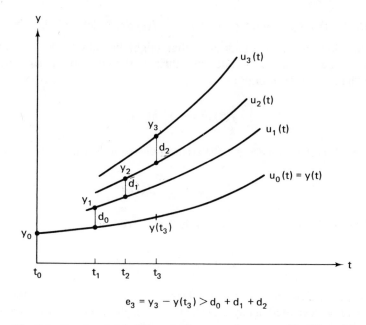

$$e_3 = y_3 - y(t_3) > d_0 + d_1 + d_2$$

Fig. 6.4. Local and global discretization errors for an unstable differential equation.

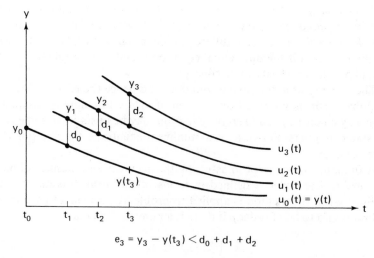

$$e_3 = y_3 - y(t_3) < d_0 + d_1 + d_2$$

Fig. 6.5. Local and global discretization errors for a stable differential equation.

is sought, but it should be independent of the step number n and the step size h_n. The above inequality can be abbreviated using "big-oh notation,"

$$d_n = O(h_n^{p+1}).$$

For example, consider Euler's method:

$$y_{n+1} = y_n + h_n f(y_n, t_n).$$

Assume the local solution $u_n(t)$ has a continuous second derivative. Then, using Taylor series near the point t_n,

$$u_n(t) = u_n(t_n) + (t - t_n)u'(t_n) + O((t - t_n)^2).$$

Using the differential equation and the initial condition defining $u_n(t)$,

$$u_n(t_{n+1}) = y_n + h_n f(y_n, t_n) + O(h_n^2).$$

Consequently,

$$d_n = y_{n+1} - u_n(t_{n+1}) = O(h_n^2).$$

We conclude that Euler's method has $p = 1$, which is first order.

Now consider the global discretization error at a fixed final point $t = t_f$. As accuracy requirements are increased, the step sizes h_n will decrease, and the total number of steps N required to reach t_f will increase. Roughly, we shall have

$$N = \frac{t_f - t_0}{h},$$

where h is the average step size. Moreover, the global error e_N can be expressed as a sum of N local errors coupled by factors describing the stability of the equation. These factors do not depend in a strong way on the step sizes, and so we can say roughly that if the local error is $O(h^{p+1})$, then the global error will be $N \cdot O(h^{p+1}) = O(h^p)$. This is why $p + 1$ was used as the exponent in the definition of order rather than p.

For Euler's method $p = 1$, so decreasing the average step size by a factor of 2 will decrease the average local error by a factor of roughly $2^{p+1} = 4$, but since about twice as many steps will be required to reach t_f, the global error will be decreased by a factor of only $2^p = 2$. With higher-order methods, the global error for smooth solutions would be reduced by a much larger factor.

It should be pointed out that in discussing numerical methods for ordinary differential equations, the word *order* can have any of several different meanings. The order of a differential equation is the index of the highest derivative appearing. For example, $y'' = t + \sin(y)$ is a second-order differential equation. The order of a system of equations sometimes refers to the number of equations in the system. For example, $y' = y + z$, $z' = y - z$ is a second-order system. The order of a numerical method for solving an ordinary differential equation is what we have been discussing here. It is the power of the step size which appears in the expression for the global error. For example, Euler's method is a first-order method.

In practice the above analysis must be modified to include roundoff errors. If a worst-case rounding error of ϵ were to occur at every step of Euler's method, then (assuming fixed step size for simplicity)

$$y_{n+1} = y_n + hf(y_n, t_n) + \epsilon,$$

and the total error due to rounding is $N\epsilon$ or $b\epsilon/h$, where $b = t_f - t_0$. Hopefully, instead of being a constant, ϵ is somewhat random with a mean of 0. In practice, the total error due to rounding usually behaves like $\epsilon\sqrt{N}$ if the machine rounds. On machines such as the IBM 360 that chop off the extra digits after arithmetic operations (this is often called truncation, and that is why we preferred *discretization error* above), the rounding error tends to grow linearly with N.

Summing the global discretization error and the roundoff error gives

$$\text{total error} \approx b\left(Ch + \frac{\epsilon}{h}\right).$$

The total error is thus minimized for some optimum value of h given approximately by

$$h_{\text{opt}} \approx \sqrt{\frac{\epsilon}{C}}.$$

The total error at the end of an integration interval is represented graphically as a function of the step size h in Fig. 6.6. As h is decreased, the discretization error decreases (linearly), and the numerical solution begins to converge to the true solution. However, when h becomes too small, the numerical solution begins to diverge because of roundoff error. The optimum value of h to use for this problem is apparently about 2^{-18}. When solving ordinary differential equations, errors due to roundoff usually take a considerable amount of computer time to build up. For example, the computation for Fig. 6.6 took about 15 minutes of Honeywell 6050 processor time. How-

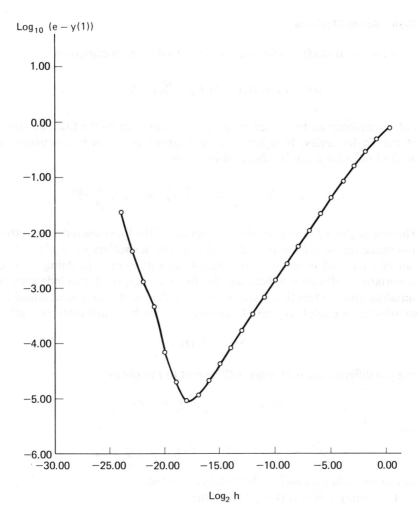

Fig. 6.6. Total error in $y(1)$ using Euler's method to solve $y' = y$, $y(0) = 1$ on a Honeywell 6050 using single-precision arithmetic ($\beta = 2$, $t = 27$).

ever, many people integrate systems of equations which require hours of computer time.

6.4. METHODS

Although there are many different step-by-step methods for numerically solving ordinary differential equations, each falls into one of four general categories.

Taylor Series Methods

If the solution $y(t)$ is smooth, it has the Taylor series expansion

$$y(t + h) = y(t) + hy'(t) + \frac{h^2}{2!}y''(t) + \ldots.$$

Euler's method can be viewed as an approximation using the first two terms of the Taylor series. If higher-order derivatives of y can be calculated, a method of order p can be obtained by using

$$y_{n+1} = y_n + hy'_n + \frac{h^2}{2!}y''_n + \frac{h^3}{3!}y'''_n + \ldots + \frac{h^p}{p!}y_n^{(p)}.$$

The first neglected term provides an estimate of the local discretization error and hence can be used to pick the step size. The derivatives $y'(x)$, $y''(x)$, etc., can be expressed in terms of the partial derivatives of f. In doing so, it is important to distinguish between the function $f(y, t)$ of two independent variables and the function $f(y(t), t)$ of one independent variable obtained by substituting the solution y into f. Starting with the basic differential equation

$$y'(t) = f(y(t), t),$$

one can differentiate both sides with respect to t to obtain

$$y'' = f_y y' + f_t = f_y \cdot f + f_t;$$

then

$$y''' = f_{yy} \cdot f^2 + f_y^2 \cdot f + f_y \cdot f_t + 2f_{ty} \cdot f + f_{tt},$$

and so on. This process is called total differentiation.

For example, if $f(y, t) = y^2 + t^2$, then

$$y' = y^2 + t^2,$$
$$y'' = 2y^3 + 2yt^2 + 2t,$$
$$y''' = 6y^4 + 8y^2t^2 + 4yt + 2t^4 + 2.$$

Notice that even though f is a polynomial in two variables, its total derivatives become more complicated as the order increases.

It is possible to write a program which computes the total derivatives for certain classes of functions f. For differential equations based on such functions, Taylor series methods can be very effective. But their general applicability is too limited to be of further interest here.

Runge-Kutta Methods

The Runge-Kutta method is designed to approximate Taylor series methods without requiring explicit definitions of, or evaluations of, derivatives beyond the first. The approximation is obtained at the expense of several evaluations of the function. The classical fourth-order Runge-Kutta method is given by

$$y_{n+1} = y_n + \tfrac{1}{6}(k_0 + 2k_1 + 2k_2 + k_3),$$

where

$$k_0 = hf(y_n, t_n),$$

$$k_1 = hf\left(y_n + \frac{1}{2}k_0, t_n + \frac{h}{2}\right),$$

$$k_2 = hf\left(y_n + \frac{1}{2}k_1, t_n + \frac{h}{2}\right).$$

$$k_3 = hf(y_n + k_2, t_n + h).$$

Notice that four evaluations of the function $f(y, t)$ are required for each step.

One may view the classical Runge-Kutta method as an extension to differential equations of Simpson's quadrature rule:

$$\int_{t_n}^{t_{n+1}} f(t)\, dt \approx \frac{h_n}{6}\left[f(t_n) + 4f\left(\frac{t_n + t_{n+1}}{2}\right) + f(t_{n+1}) \right].$$

If $f(y, t)$ is a function of t only, the two are identical.

As $h \rightarrow 0$, the classical fourth-order Runge-Kutta method agrees with the Taylor series asymptotically through terms of order h^4. However, no estimate of the local discretization error is readily available to help pick the step size. In Section 6.8 we shall describe in detail a modern modification of the Runge-Kutta method which does include step size control. Runge-Kutta methods have several advantages. They are easy to code and are numerically stable for a large class of problems. Since only one solution value y_n is needed to calculate y_{n+1}, the method is *self-starting*. The step size h_n may be changed at any step in the calculation.

Multistep Methods

In the methods considered so far the value of y_{n+1} is calculated by means of a function which depends only on t_n, y_n, and the step size h_n. It seems reasonable that more accuracy might be obtained by using information at previous points, namely $y_{n-1}, y_{n-2}, \ldots$ and $f_{n-1}, f_{n-2}, \ldots$, where $f_i = f(y_i, t_i)$.

Multistep methods based on this idea are very effective. For high accuracy they usually require less work than one-step methods, and an estimate of the discretization error can often be trivially obtained. If properly programmed, a multistep method can give output at arbitrary points efficiently without changing the value of h. The order of the method can be selected automatically and changed dynamically, thus providing methods which work on a very wide range of problems. Certain kinds of errors can be detected during the solution of the problem. Stiff equations (to be discussed later) can be handled by some multistep methods, and equations can be classified as stiff or nonstiff automatically. These advantages are obtained at the expense of an increase in program complexity and, in some cases, the possibility of numerical instability.

Linear multistep methods can be considered as special cases of the formula

$$y_{n+1} = \sum_{i=1}^{k} \alpha_i y_{n+1-i} + h \sum_{i=0}^{k} \beta_i f_{n+1-i},$$

where k is a fixed integer and either α_k or β_k is not zero. This formula defines the *general linear k-step method*. The method is called linear because each f_i enters linearly; f may or may not be a linear function of its arguments.

After the method is "started," each step requires the calculation of y_{n+1} from the known values: $y_n, y_{n-1}, \ldots, y_{n-k+1}, f_n, f_{n-1}, \ldots, f_{n-k+1}$. If $\beta_0 = 0$, the method is an *explicit* method, and the calculation is straightforward. When $\beta_0 \neq 0$, the method is an *implicit* method because f_{n+1} is needed to solve for y_{n+1}. The greater difficulty in using implicit methods is justified by their more desirable properties.

Usually two multistep methods are used together in computing each step of a solution. An explicit method called a *predictor* is followed by one or more applications of an implicit method called a *corrector*—hence the name *predictor-corrector* methods.

An example of a good predictor-corrector method is the fourth-order Adams method:

$$\text{predictor:} \quad y_{n+1} = y_n + \frac{h}{24}(55f_n - 59f_{n-1} + 37f_{n-2} - 9f_{n-3}),$$

$$\text{corrector:} \quad y_{n+1} = y_n + \frac{h}{24}(9f_{n+1} + 19f_n - 5f_{n-1} + f_{n-2}).$$

Both formulas are fourth order. They are particular instances of a family of predictors and correctors with variable order. A predictor-corrector scheme for calculating y_{n+1} is as follows:

1. Use the predictor to calculate $y_{n+1}^{(0)}$, an initial approximation to y_{n+1}. Set $i = 0$.

2. Evaluate the derivative function and set $f_{n+1}^{(i)} = f(y_{n+1}^{(i)}, t_{n+1})$.
3. Calculate a better approximation $y_{n+1}^{(i+1)}$ using the corrector formula with $f_{n+1} = f_{n+1}^{(i)}$.
4. If $|y_{n+1}^{(i+1)} - y_{n+1}^{(i)}| >$ a convergence tolerance, then increment i by 1 and go to step 2; otherwise, set $y_{n+1} = y_{n+1}^{(i+1)}$.

There are three separate processes occurring: the prediction step 1, which we shall call P; the evaluation step 2, called E; and the correction step 3, called C. Step 4 may be replaced by a step that requires exactly m iterations of the corrector. The resulting method is referred to as a P(EC)^m method. There is considerable evidence that a final evaluation to provide a more accurate f for the next time step is worthwhile; such methods are referred to as P(EC)^{m}E. If the order of the predictor is at least one less than the order of the corrector, then correcting more than twice has little effect on the result. For the Adams methods, if both stability and discretization errors are considered, it is known that P(EC)^m is not so good as P(EC)$^{m-1}$E and that P(EC)^{1}E is stable for larger step sizes than is P(EC)^{2}E. Thus, the most sensible predictor-corrector scheme for the Adams methods seems to be PECE.

Extrapolation Methods

The predictor formulas discussed in the previous section can be viewed as extrapolation methods where y_{n+1} is extrapolated from known previous values of y_i and f_i. A method which uses a different kind of extrapolation was suggested by J. A. Gaunt in 1927 and developed by W. B. Gragg in his doctoral dissertation of 1964. We shall briefly outline the computational method. There are some subtle details for which we refer the reader to a well-known paper by Bulirsch and Stoer (1966).

A one-step method is used to integrate the differential equation over a particular interval, say $0 \leq t \leq T$, for various step sizes h_j, to yield the results $Y(h_j)$. This gives a discretized approximation to a function $Y(h)$, where $Y(0) = y(T)$. An interpolating polynomial or rational function $\hat{Y}(h)$ is constructed through the calculated values. Then $Y(0)$ is usually very accurately approximated by the extrapolated value $\hat{Y}(0)$. This is another example of Richardson's deferred approach to the limit, which was mentioned in Chapter 5.

6.5. STIFF EQUATIONS

The time constant of a solution to a differential equation is the time it takes to decay by a factor of $1/e$. For example, the equation $y' = -\lambda y$ has the solution $ce^{-\lambda t}$. If λ is positive, then y decays by a factor of $1/e$ in time $1/\lambda$. A stable differential equation is called *stiff* when it has a decaying exponential particular solution with a time constant which is very small relative to the interval over which it is being solved. For the system

$$\mathbf{y}' = \mathbf{f}(\mathbf{y})$$

the decay rates are related, at least locally, to the eigenvalues of the matrix $\partial \mathbf{f}/\partial \mathbf{y} = (\partial f_i/\partial y_j)$. Consider, for example,

$$u' = 998u + 1998v,$$
$$v' = -999u - 1999v.$$

The eigenvalues of the matrix of coefficients

$$\begin{pmatrix} 998 & 1998 \\ -999 & -1999 \end{pmatrix}$$

are -1 and -1000. If $u(0) = v(0) = 1$, the solution is

$$u = 4e^{-t} - 3e^{-1000t},$$
$$v = -2e^{-t} + 3e^{-1000t},$$

as shown in Fig. 6.7. After a very short time the solution is closely approximated by

$$u = 4e^{-t},$$
$$v = -2e^{-t}.$$

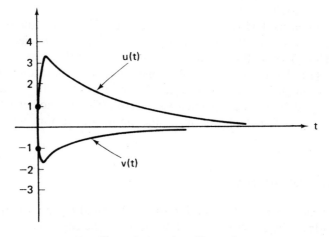

Fig. 6.7. Solution of a stiff equation.

Now suppose we were to use Euler's method to solve this system. The discretized solution may be written

$$u_{n+1} = u_n + h(998u_n + 1998v_n),$$
$$v_{n+1} = v_n + h(-999u_n - 1999v_n),$$

where $u_0 = v_0 = 1$. If we choose $h = 0.01$, then

$$u_1 = 1 + 0.01(998 + 1998) = 30.96,$$
$$v_1 = 1 + 0.01(-999 - 1999) = -28.98.$$

Try carrying the integration a few more steps to see how disastrous the results become. The problem can be corrected, temporarily at least, by using a smaller value of h. Eventually, however, roundoff and discretization errors will accumulate enough to result in another instability. This phenomenon can be visualized by considering the family of solutions associated with $u(t)$, as shown in Fig. 6.8. A transient part of the solution, which has long since decayed, prevents an increase in step size and limits the range of independent variable for which a solution can be calculated. This is particularly discon-

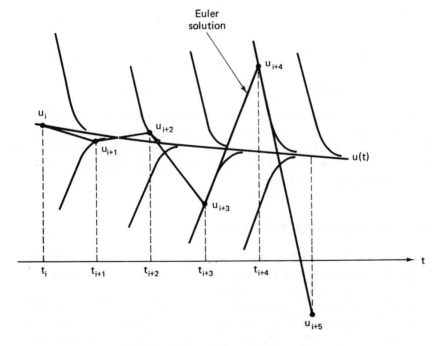

Fig. 6.8. Euler solution of stiff equation.

certing because the solution is very smooth at this point of the calculation, and one would like to increase the step size.

Most standard methods are not well suited for solving stiff equations. Consequently, as the interval h is automatically reduced and reduced, a great deal of money is often spent on a solution before the equation is recognized as being stiff. Fortunately, special methods have been devised that are very effective. Perhaps the simplest method is an implicit formula known as the *backward Euler method*:

$$y_{n+1} = y_n + hf(y_{n+1}, t_{n+1}).$$

The effect of the backward Euler method is illustrated in Fig. 6.9.

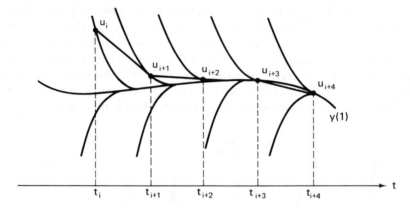

Fig. 6.9.   Backward Euler solution of stiff equation.

Developing effective methods for stiff equations is an area of active research. A discussion of the currently available methods is beyond the scope of this book and would become out of date too rapidly. The reader is advised to consult the book by Gear (1971) and recent literature in numerical analysis and mathematical software.

6.6. BOUNDARY-VALUE PROBLEMS

Thus far we have discussed only the initial value problem in which values of the dependent variables are specified for one value of the independent variable t. In some problems one wishes to solve a system of ordinary differential equations subject to values of the dependent variable which are specified at different values of t. Such a problem is called a *boundary-value problem*. Methods for solving boundary-value problems are generally very different from the methods for initial value problems, see Keller (1968). However, a popular method for solving boundary-value problems which reduces the problem to a sequence of initial value problems will be mentioned briefly.

Suppose the solution to the system

$$y' = f(y, t),$$

where

$$y = \begin{pmatrix} y_1 \\ y_2 \end{pmatrix},$$

is desired, subject to the conditions $y_1(0) = \alpha$ and $y_2(b) = \beta$, $b \neq 0$. The following iterative method may be used:

1. Choose a value ξ to approximate $y_2(0)$.
2. Solve the initial value problem:

$$\hat{y}' = f(\hat{y}, t), \quad \hat{y}(0) = \begin{pmatrix} \alpha \\ \xi \end{pmatrix},$$

3. If $|\hat{y}_2(b) - \beta|$ is small, then let $y(t) \approx \hat{y}(t)$; otherwise adjust ξ somehow (see Chapter 7) and go to step 2.

This method of solving boundary-value problems is called the *shooting method*, an appropriate name.

6.7. CHOICE OF A SUBROUTINE

It is difficult to design one program to solve all initial value problems for ordinary differential equations efficiently. Among other things, consideration must be given to special properties of particular problems, to stiffness, to the complexity and cost of evaluating equations, to the expected accuracy requirements, to the types of output desired, and to the ease with which the program can be read, used, and, if necessary, modified by its users.

Of the many available subroutines, we have chosen to include a Fortran subroutine RKF45, designed and written by H. A. Watts and L. F. Shampine. It was not an easy choice to make, and, in fact, two preliminary drafts of this book used two other subroutines. In this section, we shall explain this particular choice and why other choices might be made in other situations.

Several recent studies have compared various methods and programs for solving initial value problems. The reader who wishes to have more detail about the choice of a subroutine should see the papers by Hull et al. (1972), Krogh (1973), Shampine et al. (1976), and Enright and Hull (1976).

It is important to distinguish between a method and a subroutine. A method can usually be described by a few fairly simple formulas. In fact, several methods were described in the previous section. A complete problem-solving subroutine, on the other hand, is a much more complex object. It must handle a multitude of details like step size selection, error control, temporary storage management, communication with other programs, detection of errors in the calling sequence, and possible discontinuities and singularities. Two different subroutines which implement the same basic method can therefore perform very differently.

The user is likely to know the accuracy needed in a solution, but he may have no idea of the step size needed to achieve it. Moreover, the step size requirements generally change during the solution. Hence, we regard it as essential that a subroutine for general use include automatic step size selection and discretization error control.

We are primarily concerned with nonstiff problems. A reasonable subroutine with automatic error control will usually work on stiff problems, but it may be quite inefficient. Some recently developed subroutines attempt to detect excessive stiffness and warn the user that another approach may be less expensive.

Taylor series methods are applicable only to special problems in which $f(y, t)$ can be specified in a symbolic, easily differentiated form. Since we want methods for which the user need only supply a subroutine for evaluating $f(y, t)$, we shall consider only Runge-Kutta, multistep, and extrapolation methods.

A single Runge-Kutta method necessarily has a fixed order. For example, the classical method is fourth order. Methods with different coefficients are necessary for different orders. Moreover, a single method will not have any way of monitoring discretization error and hence cannot select step size. It is necessary to combine two different methods to achieve automatic step size control. Runge-Kutta methods which do this in an efficient way have been developed only recently. These combined methods are still of fixed order and hence are effective only in a limited range of accuracy requirements.

A well-written subroutine based on Runge-Kutta methods can be fairly easy to read, understand, use, and modify. These are the main reasons we choose such a subroutine for this book. The subroutine RKF45 described in the next section is the best general-purpose implementation of Runge-Kutta methods we know of.

The Adams methods are a family of multistep methods of varying order. A well-written variable-order, variable-step Adams subroutine will generally be more efficient over a wide range of accuracy requirements than a fixed-order, variable-step Runge-Kutta subroutine. Examples of such subroutines include DVDQ and DVOA, written by Fred Krogh of Caltech's Jet Propulsion Laboratory, DIFSUB by C. W. Gear (1971), STEP and DE by L. F. Shampine and M. K. Gordon (1975), and VOAS by A. Sedgwick (1973). Because of the variable order, all these subroutines tend to be more complicated. Hence, they require more storage and are somewhat harder to read and understand than RKF45. Moreover, at modest accuracy requests (local error less than 10^{-5} or 10^{-6}) RKF45 is just as efficient as the more sophisticated Adams methods.

The subroutines currently available which use extrapolation methods are somewhat less flexible than those based on the other methods. Moreover, they tend to be somewhat less efficient, especially for high accuracy requirements or if output results are desired throughout the interval rather than just at the end. Work on implementation of extrapolation methods continues, however, and so they should always be considered in any comparison.

6.8. SUBROUTINE RKF45

RKF45 is a subroutine for solving initial value problems in ordinary differential equations which is based on Runge-Kutta formulas developed by E. Fehlberg in 1970 and implemented by L. F. Shampine and H. A. Watts in 1974. It requires six function evaluations per step. Four of these function values are combined with one set of coefficients to produce a fourth-order method, and all six values are combined with another set of coefficients to produce a fifth-order method. Comparison of the two values yields an error estimate which is used for step size control.

We shall not attempt to derive the complete formulas in detail but instead shall investigate a simpler, second-order method. This method uses only two function evaluations per step. The first is

$$k_1 = h_n f(y_n, t_n).$$

Then a fractional step based on k_1 is taken. We introduce two as yet unknown coefficients α and β. Let

$$k_2 = h_n f(y_n + \beta k_1, t_n + \alpha h_n).$$

Finally, the two function values are combined to make a complete step, involving two more unknown coefficients:

$$y_{n+1} = y_n + \gamma_1 k_1 + \gamma_2 k_2.$$

To determine the coefficients, k_1 and k_2 are expanded in a Taylor series (in two variables) about (t_n, y_n), giving

$$k_1 = h_n f_n,$$
$$k_2 = h_n(f_n + \beta k_1 f_{y,n} + \alpha h_n f_{t,n} + \ldots),$$
$$y_{n+1} = y_n + (\gamma_1 + \gamma_2)h_n f_n + \gamma_2 \beta h_n^2 f_{y,n} f_n + \gamma_2 \alpha h_n^2 f_{t,n} + \ldots.$$

The expansion of y_{n+1} can then be compared with the expansion for the true *local* solution, $u_n(t)$, defined in Section 6.3:

$$u_n(t_{n+1}) = u_n(t_n) + h_n u_n'(t_n) + \frac{h_n^2}{2} u_n''(t_n) + \ldots$$

$$= y_n + h_n f_n + \frac{h_n^2}{2}(f_{y,n} f_n + f_{t,n}) + \ldots.$$

Matching coefficients of powers h_n leads to three equations involving the four unknown coefficients:

$$\gamma_1 + \gamma_2 = 1,$$

$$\gamma_2 \beta = \frac{1}{2},$$

$$\gamma_2 \alpha = \frac{1}{2}.$$

Picking one of these coefficients, say α, as a parameter produces a one-parameter family of Runge-Kutta methods:

$$k_1 = h_n f(y_n, t_n),$$

$$k_2 = h_n f(y_n + \alpha k_1, t_n + \alpha h_n),$$

$$y_{n+1} = y_n + \left(1 - \frac{1}{2\alpha}\right) k_1 + \frac{1}{2\alpha} k_2.$$

Two obvious choices of α are $\frac{1}{2}$ and 1. These define methods closely related to the rectangle rule and the trapezoid rule, respectively. Examination of the first terms neglected in the above expansions will reveal that no choice of α can eliminate these terms for all functions $f(y, t)$, and so the method is second order for any α.

The formulas actually used by RKF45 involve six function evaluations per step, k_1 through k_6. These are defined by

$$k_i = h_n f(y_n + \sum_{j=1}^{i-1} \beta_{ij} k_j, t_n + \alpha_i h_n), \qquad i = 1, \ldots, 6.$$

The new value y_{n+1} is then obtained using a weighted combination of the 6 k's:

$$y_{n+1} = y_n + \sum_{i=1}^{6} \gamma_i k_i.$$

There are 27 coefficients, 6 α's, 15 β's, and 6 γ's. Notice that the β's form a lower triangular array so that each k_i is obtained from previous k's.

To determine the coefficients, all the k's can be expanded in Taylor series, the expansions substituted into the formula for y_{n+1}, and the result compared with the Taylor series for $u_n(t_{n+1})$. It is possible to find values of the coefficients so that the difference between two expansions starts with a term including h_n^6 but not h_n^5. The resulting method is consequently of fifth order. This is somewhat of a surprise, because for $p = 1, 2, 3$, and 4 it is possible to obtain a pth-order method with p function evaluations, while for $p = 5$ or 6, p function evaluations will produce only a $(p - 1)$st-order method. (This partially accounts for the popularity of the classical fourth-order method mentioned in the last section; it takes two more function evaluations to obtain one more order of accuracy.)

There are many possible choices of the coefficients that produce fifth-order methods. The values discovered by Fehlberg are given in Table 6.1.

Table 6.1

α_i	β_{ij}					γ_i	γ_i^*
0						$\frac{16}{135}$	$\frac{25}{216}$
$\frac{1}{4}$	$\frac{1}{4}$					0	0
$\frac{3}{8}$	$\frac{3}{32}$	$\frac{9}{32}$				$\frac{6656}{12825}$	$\frac{1408}{2565}$
$\frac{12}{13}$	$\frac{1932}{2197}$	$-\frac{7200}{2197}$	$\frac{7296}{2197}$			$\frac{28561}{56430}$	$\frac{2197}{4104}$
1	$\frac{439}{216}$	-8	$\frac{3680}{513}$	$-\frac{845}{4104}$		$-\frac{9}{50}$	$-\frac{1}{5}$
$\frac{1}{2}$	$-\frac{8}{27}$	2	$-\frac{3544}{2565}$	$\frac{1859}{4104}$	$-\frac{11}{40}$	$\frac{2}{55}$	0

The "extra" function evaluation is not wasted, however. Fehlberg also found a set of coefficients γ_i^*, so that four of the six k's can be combined to give a second value y_{n+1}^* which is accurate to fourth order. That is,

$$y_{n+1}^* = y_n + \sum_{i=1}^{6} \gamma_i^* k_i.$$

These coefficients are also given in the table. The program does not actually compute y_{n+1}^* but instead computes the error estimate $\sum_{i=1}^{6} (\gamma_i - \gamma_i^*) k_i$ to use for step size control.

The sample main program at the end of this section uses RKF45 to find the motion of two bodies under mutual gravitational attraction. Let $x(t)$ and

$y(t)$ denote the position of one body in a coordinate system with the origin fixed in the other body. The differential equations derived from Newton's laws of motion are

$$x''(t) = \frac{-\alpha^2 x(t)}{R(t)}$$

$$y''(t) = \frac{-\alpha^2 y(t)}{R(t)},$$

where

$$R(t) = [x(t)^2 + y(t)^2]^{3/2}$$

and α is a constant involving the gravitational constant, the masses of the two bodies, and the units of measurement.

If the initial conditions are chosen as

$$x(0) = 1 - e, \qquad x'(0) = 0,$$

$$y(0) = 0, \qquad y'(0) = \alpha \left(\frac{1+e}{1-e}\right)^{1/2},$$

for some parameter e with $0 \le e < 1$, then the solution turns out to be periodic with period $2\pi/\alpha$. The orbit is an ellipse with eccentricity e and with one focus at the origin.

To write this as a system of four first-order equations, we introduce

$$y_1 = x, \qquad y_2 = y, \qquad y_3 = x', \qquad y_4 = y'.$$

The equations and initial conditions then become

$$R = \frac{(y_1^2 + y_2^2)^{3/2}}{\alpha^2},$$

$$y_1' = y_3, \qquad y_1(0) = 1 - e,$$
$$y_2' = y_4, \qquad y_2(0) = 0,$$
$$y_3' = -\frac{y_1}{R}, \qquad y_3(0) = 0,$$
$$y_4' = -\frac{y_2}{R}, \qquad y_4(0) = \alpha \left(\frac{1+e}{1-e}\right)^{1/2}.$$

By rescaling the time variable, it is possible to eliminate α, but we have not done this because we wish to illustrate the use of Fortran Common to pass parameters such as α from the main program to the subroutine defining the equations.

The parameter IFLAG is an important control variable. It should be set to 1 for the first entry to RKF45. Ordinarily, RKF45 will reset it to 2, and it should be left at 2 for subsequent entries. Values other than 2 returned by RKF45 signal various warning and error conditions described in detail in the comments. IFLAG = 4 and IFLAG = 7 are warnings that RKF45 must work very hard to obtain the requested accuracy. It is possible to continue, but the user may want to consider increasing the error tolerances or changing to a subroutine which uses a multistep method. IFLAG = 3 indicates that too much relative accuracy is being requested, and IFLAG = 5 or 6 indicates that the error tolerances must be changed before continuing. IFLAG = 8 indicates that RKF45 is being called incorrectly. The user is strongly advised to include a check on IFLAG in his main program.

In this sample run, we have taken $e = 0.25$ and $\alpha = \pi/4$ and have printed the position for $0 \le t \le 12$ in steps of 0.5. The output is in Table 6.2. Notice that the orbit is periodic with a period of $t = 8$.

Table 6.2   OUTPUT FROM SAMPLE PROGRAM

0.0	0.750000000	0.000000000
0.5	0.619768032	0.477791373
1.0	0.294417538	0.812178519
1.5	−0.105176382	0.958038092
2.0	−0.490299793	0.939874996
2.5	−0.813942832	0.799590802
3.0	−1.054031517	0.575706078
3.5	−1.200735042	0.300160708
4.0	−1.250000001	−0.000000001
4.5	−1.200735042	−0.300160709
5.0	−1.054031517	−0.575706079
5.5	−0.813942832	−0.799590803
6.0	−0.490299793	−0.939874996
6.5	−0.105176383	−0.958038092
7.0	0.294417537	−0.812178518
7.5	0.619768031	−0.477791370
8.0	0.749999996	0.000000006
8.5	0.619768024	0.477791379
9.0	0.294417526	0.812178522
9.5	−0.105176395	0.958038091
10.0	−0.490299806	0.939874991
10.5	−0.813942843	0.799590794
11.0	−1.054031524	0.575706068
11.5	−1.200735047	0.300160697
12.0	−1.250000002	−0.000000011

```
C  SAMPLE PROGRAM FOR RKF45
C
      SUBROUTINE ORBIT (T, Y, YP)
      REAL T, Y(4), YP(4), R, ALFASQ
      COMMON ALFASQ
      R = Y(1)*Y(1) + Y(2)*Y(2)
      R = R*SQRT(R)/ALFASQ
      YP(1) = Y(3)
      YP(2) = Y(4)
      YP(3) = -Y(1)/R
      YP(4) = -Y(2)/R
      RETURN
      END
C
      EXTERNAL ORBIT
      REAL T,Y(4),TOUT,RELERR,ABSERR
      REAL TFINAL,TPRINT,ECC,ALFA,ALFASQ,WORK(27)
      INTEGER IWORK(5), IFLAG, NEQN
      COMMON ALFASQ
      ECC = 0.25
      ALFA = 3.141592653589/4.0
      ALFASQ = ALFA*ALFA
      NEQN = 4
      T = 0.0
      Y(1) = 1.0 - ECC
      Y(2) = 0.0
      Y(3) = 0.0
      Y(4) = ALFA*SQRT( (1.0 + ECC)/(1.0 - ECC) )
      RELERR = 1.0E-9
      ABSERR = 0.0
      TFINAL = 12.0
      TPRINT = 0.5
      IFLAG = 1
      TOUT = T
   10 CALL RKF45(ORBIT,NEQN,Y,T,TOUT,RELERR,ABSERR,IFLAG,WORK,IWORK)
      WRITE(6,11) T, Y(1), Y(2)
      GO TO (80,20,30,40,50,60,70,80), IFLAG
   20 TOUT = T + TPRINT
      IF (T .LT. TFINAL) GO TO 10
      STOP
   30 WRITE(6,31) RELERR,ABSERR
      GO TO 10
   40 WRITE(6,41)
      GO TO 10
   50 ABSERR = 1.0E-9
      WRITE(6,31) RELERR,ABSERR
      GO TO 10
   60 RELERR = 10.0*RELERR
      WRITE(6,31) RELERR,ABSERR
      IFLAG = 2
      GO TO 10
   70 WRITE(6,71)
      IFLAG = 2
      GO TO 10
   80 WRITE(6,81)
      STOP
C
   11 FORMAT(F5.1, 2F15.9)
   31 FORMAT(17H TOLERANCES RESET, 2E12.3)
   41 FORMAT(11H MANY STEPS)
   71 FORMAT(12H MUCH OUTPUT)
   81 FORMAT(14H IMPROPER CALL)
      END
```

```
      SUBROUTINE RKF45(F,NEQN,Y,T,TOUT,RELERR,ABSERR,IFLAG,WORK,IWORK)
C
C     FEHLBERG FOURTH-FIFTH ORDER RUNGE-KUTTA METHOD
C
C     WRITTEN BY H.A.WATTS AND L.F.SHAMPINE
C                SANDIA LABORATORIES
C                ALBUQUERQUE,NEW MEXICO
C
C     RKF45 IS PRIMARILY DESIGNED TO SOLVE NON-STIFF AND MILDLY STIFF
C     DIFFERENTIAL EQUATIONS WHEN DERIVATIVE EVALUATIONS ARE INEXPENSIVE.
C     RKF45 SHOULD GENERALLY NOT BE USED WHEN THE USER IS DEMANDING
C     HIGH ACCURACY.
C
C ABSTRACT
C
C     SUBROUTINE  RKF45  INTEGRATES A SYSTEM OF NEQN FIRST ORDER
C     ORDINARY DIFFERENTIAL EQUATIONS OF THE FORM
C             DY(I)/DT = F(T,Y(1),Y(2),...,Y(NEQN))
C               WHERE THE Y(I) ARE GIVEN AT T .
C     TYPICALLY THE SUBROUTINE IS USED TO INTEGRATE FROM T TO TOUT BUT IT
C     CAN BE USED AS A ONE-STEP INTEGRATOR TO ADVANCE THE SOLUTION A
C     SINGLE STEP IN THE DIRECTION OF TOUT.  ON RETURN THE PARAMETERS IN
C     THE CALL LIST ARE SET FOR CONTINUING THE INTEGRATION. THE USER HAS
C     ONLY TO CALL RKF45 AGAIN (AND PERHAPS DEFINE A NEW VALUE FOR TOUT).
C     ACTUALLY, RKF45 IS AN INTERFACING ROUTINE WHICH CALLS SUBROUTINE
C     RKFS FOR THE SOLUTION.  RKFS IN TURN CALLS SUBROUTINE  FEHL WHICH
C     COMPUTES AN APPROXIMATE SOLUTION OVER ONE STEP.
C
C     RKF45  USES THE RUNGE-KUTTA-FEHLBERG (4,5)  METHOD DESCRIBED
C     IN THE REFERENCE
C     E.FEHLBERG , LOW-ORDER CLASSICAL RUNGE-KUTTA FORMULAS WITH STEPSIZE
C              CONTROL , NASA TR R-315
C
C     THE PERFORMANCE OF RKF45 IS ILLUSTRATED IN THE REFERENCE
C     L.F.SHAMPINE,H.A.WATTS,S.DAVENPORT, SOLVING NON-STIFF ORDINARY
C              DIFFERENTIAL EQUATIONS-THE STATE OF THE ART ,
C              SANDIA LABORATORIES REPORT SAND75-0182 ,
C              TO APPEAR IN SIAM REVIEW.
C
C
C     THE PARAMETERS REPRESENT-
C       F -- SUBROUTINE F(T,Y,YP) TO EVALUATE DERIVATIVES YP(I)=DY(I)/DT
C       NEQN -- NUMBER OF EQUATIONS TO BE INTEGRATED
C       Y(*) -- SOLUTION VECTOR AT T
C       T -- INDEPENDENT VARIABLE
C       TOUT -- OUTPUT POINT AT WHICH SOLUTION IS DESIRED
C       RELERR,ABSERR -- RELATIVE AND ABSOLUTE ERROR TOLERANCES FOR LOCAL
C            ERROR TEST. AT EACH STEP THE CODE REQUIRES THAT
C                 ABS(LOCAL ERROR) .LE. RELERR*ABS(Y) + ABSERR
C            FOR EACH COMPONENT OF THE LOCAL ERROR AND SOLUTION VECTORS
C       IFLAG -- INDICATOR FOR STATUS OF INTEGRATION
C       WORK(*) -- ARRAY TO HOLD INFORMATION INTERNAL TO RKF45 WHICH IS
C            NECESSARY FOR SUBSEQUENT CALLS. MUST BE DIMENSIONED
C            AT LEAST  3+6*NEQN
```

```
C      IWORK(*) -- INTEGER ARRAY USED TO HOLD INFORMATION INTERNAL TO
C           RKF45 WHICH IS NECESSARY FOR SUBSEQUENT CALLS. MUST BE
C           DIMENSIONED AT LEAST  5
C
C
C  FIRST CALL TO RKF45
C
C    THE USER MUST PROVIDE STORAGE IN HIS CALLING PROGRAM FOR THE ARRAYS
C    IN THE CALL LIST       Y(NEQN) , WORK(3+6*NEQN) , IWORK(5)  ,
C    DECLARE F IN AN EXTERNAL STATEMENT, SUPPLY SUBROUTINE F(T,Y,YP) AND
C    INITIALIZE THE FOLLOWING PARAMETERS-
C
C      NEQN -- NUMBER OF EQUATIONS TO BE INTEGRATED. (NEQN .GE. 1)
C      Y(*) -- VECTOR OF INITIAL CONDITIONS
C      T -- STARTING POINT OF INTEGRATION , MUST BE A VARIABLE
C      TOUT -- OUTPUT POINT AT WHICH SOLUTION IS DESIRED.
C           T=TOUT IS ALLOWED ON THE FIRST CALL ONLY, IN WHICH CASE
C           RKF45 RETURNS WITH IFLAG=2 IF CONTINUATION IS POSSIBLE.
C      RELERR,ABSERR -- RELATIVE AND ABSOLUTE LOCAL ERROR TOLERANCES
C           WHICH MUST BE NON-NEGATIVE. RELERR MUST BE A VARIABLE WHILE
C           ABSERR MAY BE A CONSTANT. THE CODE SHOULD NORMALLY NOT BE
C           USED WITH RELATIVE ERROR CONTROL SMALLER THAN ABOUT 1.E-8 .
C           TO AVOID LIMITING PRECISION DIFFICULTIES THE CODE REQUIRES
C           RELERR TO BE LARGER THAN AN INTERNALLY COMPUTED RELATIVE
C           ERROR PARAMETER WHICH IS MACHINE DEPENDENT. IN PARTICULAR,
C           PURE ABSOLUTE ERROR IS NOT PERMITTED. IF A SMALLER THAN
C           ALLOWABLE VALUE OF RELERR IS ATTEMPTED, RKF45 INCREASES
C           RELERR APPROPRIATELY AND RETURNS CONTROL TO THE USER BEFORE
C           CONTINUING THE INTEGRATION.
C      IFLAG -- +1,-1 INDICATOR TO INITIALIZE THE CODE FOR EACH NEW
C           PROBLEM. NORMAL INPUT IS +1. THE USER SHOULD SET IFLAG=-1
C           ONLY WHEN ONE-STEP INTEGRATOR CONTROL IS ESSENTIAL. IN THIS
C           CASE, RKF45 ATTEMPTS TO ADVANCE THE SOLUTION A SINGLE STEP
C           IN THE DIRECTION OF TOUT EACH TIME IT IS CALLED. SINCE THIS
C           MODE OF OPERATION RESULTS IN EXTRA COMPUTING OVERHEAD, IT
C           SHOULD BE AVOIDED UNLESS NEEDED.
C
C
C  OUTPUT FROM RKF45
C
C      Y(*) -- SOLUTION AT T
C      T -- LAST POINT REACHED IN INTEGRATION.
C      IFLAG = 2 -- INTEGRATION REACHED TOUT. INDICATES SUCCESSFUL RETURN
C                AND IS THE NORMAL MODE FOR CONTINUING INTEGRATION.
C            =-2 -- A SINGLE SUCCESSFUL STEP IN THE DIRECTION OF TOUT
C                HAS BEEN TAKEN. NORMAL MODE FOR CONTINUING
C                INTEGRATION ONE STEP AT A TIME.
C            = 3 -- INTEGRATION WAS NOT COMPLETED BECAUSE RELATIVE ERROR
C                TOLERANCE WAS TOO SMALL. RELERR HAS BEEN INCREASED
C                APPROPRIATELY FOR CONTINUING.
C            = 4 -- INTEGRATION WAS NOT COMPLETED BECAUSE MORE THAN
C                3000 DERIVATIVE EVALUATIONS WERE NEEDED. THIS
C                IS APPROXIMATELY 500 STEPS.
```

```
C            = 5 -- INTEGRATION WAS NOT COMPLETED BECAUSE SOLUTION
C                   VANISHED MAKING A PURE RELATIVE ERROR TEST
C                   IMPOSSIBLE. MUST USE NON-ZERO ABSERR TO CONTINUE.
C                   USING THE ONE-STEP INTEGRATION MODE FOR ONE STEP
C                   IS A GOOD WAY TO PROCEED.
C            = 6 -- INTEGRATION WAS NOT COMPLETED BECAUSE REQUESTED
C                   ACCURACY COULD NOT BE ACHIEVED USING SMALLEST
C                   ALLOWABLE STEPSIZE. USER MUST INCREASE THE ERROR
C                   TOLERANCE BEFORE CONTINUED INTEGRATION CAN BE
C                   ATTEMPTED.
C            = 7 -- IT IS LIKELY THAT RKF45 IS INEFFICIENT FOR SOLVING
C                   THIS PROBLEM. TOO MUCH OUTPUT IS RESTRICTING THE
C                   NATURAL STEPSIZE CHOICE. USE THE ONE-STEP INTEGRATOR
C                   MODE.
C            = 8 -- INVALID INPUT PARAMETERS
C                   THIS INDICATOR OCCURS IF ANY OF THE FOLLOWING IS
C                   SATISFIED -   NEQN .LE. 0
C                             T=TOUT  AND  IFLAG .NE. +1 OR -1
C                             RELERR OR ABSERR .LT. 0.
C                             IFLAG .EQ. 0  OR  .LT. -2  OR  .GT. 8
C       WORK(*),IWORK(*) -- INFORMATION WHICH IS USUALLY OF NO INTEREST
C            TO THE USER BUT NECESSARY FOR SUBSEQUENT CALLS.
C            WORK(1),...,WORK(NEQN) CONTAIN THE FIRST DERIVATIVES
C            OF THE SOLUTION VECTOR Y AT T. WORK(NEQN+1) CONTAINS
C            THE STEPSIZE H TO BE ATTEMPTED ON THE NEXT STEP.
C            IWORK(1) CONTAINS THE DERIVATIVE EVALUATION COUNTER.
C
C
C  SUBSEQUENT CALLS TO RKF45
C
C     SUBROUTINE RKF45 RETURNS WITH ALL INFORMATION NEEDED TO CONTINUE
C     THE INTEGRATION. IF THE INTEGRATION REACHED TOUT, THE USER NEED ONLY
C     DEFINE A NEW TOUT AND CALL RKF45 AGAIN. IN THE ONE-STEP INTEGRATOR
C     MODE (IFLAG=-2) THE USER MUST KEEP IN MIND THAT EACH STEP TAKEN IS
C     IN THE DIRECTION OF THE CURRENT TOUT. UPON REACHING TOUT (INDICATED
C     BY CHANGING IFLAG TO 2),THE USER MUST THEN DEFINE A NEW TOUT AND
C     RESET IFLAG TO -2 TO CONTINUE IN THE ONE-STEP INTEGRATOR MODE.
C
C     IF THE INTEGRATION WAS NOT COMPLETED BUT THE USER STILL WANTS TO
C     CONTINUE (IFLAG=3,4 CASES), HE JUST CALLS RKF45 AGAIN. WITH IFLAG=3
C     THE RELERR PARAMETER HAS BEEN ADJUSTED APPROPRIATELY FOR CONTINUING
C     THE INTEGRATION. IN THE CASE OF IFLAG=4 THE FUNCTION COUNTER WILL
C     BE RESET TO 0 AND ANOTHER 3000 FUNCTION EVALUATIONS ARE ALLOWED.
C
C     HOWEVER,IN THE CASE IFLAG=5, THE USER MUST FIRST ALTER THE ERROR
C     CRITERION TO USE A POSITIVE VALUE OF ABSERR BEFORE INTEGRATION CAN
C     PROCEED. IF HE DOES NOT,EXECUTION IS TERMINATED.
C
C     ALSO,IN THE CASE IFLAG=6, IT IS NECESSARY FOR THE USER TO RESET
C     IFLAG TO 2 (OR -2 WHEN THE ONE-STEP INTEGRATION MODE IS BEING USED)
C     AS WELL AS INCREASING EITHER ABSERR,RELERR OR BOTH BEFORE THE
C     INTEGRATION CAN BE CONTINUED. IF THIS IS NOT DONE, EXECUTION WILL
C     BE TERMINATED. THE OCCURRENCE OF IFLAG=6 INDICATES A TROUBLE SPOT
C     (SOLUTION IS CHANGING RAPIDLY,SINGULARITY MAY BE PRESENT) AND IT
C     OFTEN IS INADVISABLE TO CONTINUE.
```

```
C
C     IF IFLAG=7 IS ENCOUNTERED, THE USER SHOULD USE THE ONE-STEP
C     INTEGRATION MODE WITH THE STEPSIZE DETERMINED BY THE CODE OR
C     CONSIDER SWITCHING TO THE ADAMS CODES DE/STEP,INTRP. IF THE USER
C     INSISTS UPON CONTINUING THE INTEGRATION WITH RKF45, HE MUST RESET
C     IFLAG TO 2 BEFORE CALLING RKF45 AGAIN. OTHERWISE,EXECUTION WILL BE
C     TERMINATED.
C
C     IF IFLAG=8 IS OBTAINED, INTEGRATION CAN NOT BE CONTINUED UNLESS
C     THE INVALID INPUT PARAMETERS ARE CORRECTED.
C
C     IT SHOULD BE NOTED THAT THE ARRAYS WORK,IWORK CONTAIN INFORMATION
C     REQUIRED FOR SUBSEQUENT INTEGRATION. ACCORDINGLY, WORK AND IWORK
C     SHOULD NOT BE ALTERED.
C
C
      INTEGER NEQN,IFLAG,IWORK(5)
      REAL Y(NEQN),T,TOUT,RELERR,ABSERR,WORK(1)
C     IF COMPILER CHECKS SUBSCRIPTS, CHANGE WORK(1) TO WORK(3+6*NEQN)
C
      EXTERNAL F
C
      INTEGER K1,K2,K3,K4,K5,K6,K1M
C
C
C     COMPUTE INDICES FOR THE SPLITTING OF THE WORK ARRAY
C
      K1M=NEQN+1
      K1=K1M+1
      K2=K1+NEQN
      K3=K2+NEQN
      K4=K3+NEQN
      K5=K4+NEQN
      K6=K5+NEQN
C
C     THIS INTERFACING ROUTINE MERELY RELIEVES THE USER OF A LONG
C     CALLING LIST VIA THE SPLITTING APART OF TWO WORKING STORAGE
C     ARRAYS. IF THIS IS NOT COMPATIBLE WITH THE USERS COMPILER,
C     HE MUST USE RKFS DIRECTLY.
C
      CALL  RKFS(F,NEQN,Y,T,TOUT,RELERR,ABSERR,IFLAG,WORK(1),WORK(K1M),
     1          WORK(K1),WORK(K2),WORK(K3),WORK(K4),WORK(K5),WORK(K6),
     2          WORK(K6+1),IWORK(1),IWORK(2),IWORK(3),IWORK(4),IWORK(5))
C
      RETURN
      END
```

```
      SUBROUTINE RKFS(F,NEQN,Y.T,TOUT,RELERR,ABSERR,IFLAG,YP,H,F1,F2,F3,
     1          F4,F5,SAVRE,SAVAE,NFE,KOP,INIT,JFLAG,KFLAG)
C
C     FEHLBERG FOURTH-FIFTH ORDER RUNGE-KUTTA METHOD
C
C
C     RKFS INTEGRATES A SYSTEM OF FIRST ORDER ORDINARY DIFFERENTIAL
C     EQUATIONS AS DESCRIBED IN THE COMMENTS FOR RKF45 .
C     THE ARRAYS YP,F1,F2,F3,F4,AND F5 (OF DIMENSION AT LEAST NEQN) AND
C     THE VARIABLES H,SAVRE,SAVAE,NFE,KOP,INIT,JFLAG,AND KFLAG ARE USED
C     INTERNALLY BY THE CODE AND APPEAR IN THE CALL LIST TO ELIMINATE
C     LOCAL RETENTION OF VARIABLES BETWEEN CALLS. ACCORDINGLY, THEY
C     SHOULD NOT BE ALTERED. ITEMS OF POSSIBLE INTEREST ARE
C        YP - DERIVATIVE OF SOLUTION VECTOR AT T
C        H  - AN APPROPRIATE STEPSIZE TO BE USED FOR THE NEXT STEP
C        NFE- COUNTER ON THE NUMBER OF DERIVATIVE FUNCTION EVALUATIONS
C
C
      LOGICAL HFAILD,OUTPUT
C
      INTEGER   NEQN,IFLAG,NFE,KOP,INIT,JFLAG,KFLAG
      REAL   Y(NEQN),T,TOUT,RELERR,ABSERR,H,YP(NEQN),
     1   F1(NEQN),F2(NEQN),F3(NEQN),F4(NEQN),F5(NEQN),SAVRE,
     2   SAVAE
C
      EXTERNAL F
C
      REAL   A,AE,DT,EE,EEOET,ESTTOL,ET,HMIN,REMIN,RER,S,
     1   SCALE,TOL,TOLN,U26,EPSP1,EPS,YPK
C
      INTEGER   K,MAXNFE,MFLAG
C
      REAL   AMAX1,AMIN1
C
C     REMIN IS THE MINIMUM ACCEPTABLE VALUE OF RELERR.  ATTEMPTS
C     TO OBTAIN HIGHER ACCURACY WITH THIS SUBROUTINE ARE USUALLY
C     VERY EXPENSIVE AND OFTEN UNSUCCESSFUL.
C
      DATA REMIN/1.E-12/
C
C
C     THE EXPENSE IS CONTROLLED BY RESTRICTING THE NUMBER
C     OF FUNCTION EVALUATIONS TO BE APPROXIMATELY MAXNFE.
C     AS SET, THIS CORRESPONDS TO ABOUT 500 STEPS.
C
      DATA MAXNFE/3000/
```

```
C
C
C     CHECK INPUT PARAMETERS
C
C
      IF (NEQN .LT. 1) GO TO 10
      IF ((RELERR .LT. 0.0)  .OR.  (ABSERR .LT. 0.0)) GO TO 10
      MFLAG=IABS(IFLAG)
      IF ((MFLAG .EQ. 0) .OR. (MFLAG .GT. 8)) GO TO 10
      IF (MFLAG .NE. 1) GO TO 20
C
C     FIRST CALL, COMPUTE MACHINE EPSILON
C
      EPS = 1.0
    5 EPS = EPS/2.0
      EPSP1 = EPS + 1.0
      IF (EPSP1 .GT. 1.0) GO TO 5
      U26 = 26.0*EPS
      GO TO 50
C
C     INVALID INPUT
   10 IFLAG=8
      RETURN
C
C     CHECK CONTINUATION POSSIBILITIES
C
   20 IF ((T .EQ. TOUT) .AND. (KFLAG .NE. 3)) GO TO 10
      IF (MFLAG .NE. 2) GO TO 25
C
C     IFLAG = +2 OR -2
      IF ((KFLAG .EQ. 3) .OR. (INIT .EQ. 0)) GO TO 45
      IF (KFLAG .EQ. 4) GO TO 40
      IF ((KFLAG .EQ. 5)  .AND.  (ABSERR .EQ. 0.0)) GO TO 30
      IF ((KFLAG .EQ. 6)  .AND.  (RELERR .LE. SAVRE)  .AND.
     1    (ABSERR .LE. SAVAE)) GO TO 30
      GO TO 50
C
C     IFLAG = 3,4,5,6,7 OR 8
   25 IF. (IFLAG .EQ. 3) GO TO 45
      IF (IFLAG .EQ. 4) GO TO 40
      IF ((IFLAG .EQ. 5) .AND. (ABSERR .GT. 0.0)) GO TO 45
```

```
C
C     INTEGRATION CANNOT BE CONTINUED SINCE USER DID NOT RESPOND TO
C     THE INSTRUCTIONS PERTAINING TO IFLAG=5,6,7 OR 8
   30 STOP
C
C     RESET FUNCTION EVALUATION COUNTER
   40 NFE=0
      IF (MFLAG .EQ. 2) GO TO 50
C
C     RESET FLAG VALUE FROM PREVIOUS CALL
   45 IFLAG=JFLAG
      IF (KFLAG .EQ. 3) MFLAG=IABS(IFLAG)
C
C     SAVE INPUT IFLAG AND SET CONTINUATION FLAG VALUE FOR SUBSEQUENT
C     INPUT CHECKING
   50 JFLAG=IFLAG
      KFLAG=0
C
C     SAVE RELERR AND ABSERR FOR CHECKING INPUT ON SUBSEQUENT CALLS
      SAVRE=RELERR
      SAVAE=ABSERR
C
C     RESTRICT RELATIVE ERROR TOLERANCE TO BE AT LEAST AS LARGE AS
C     2*EPS+REMIN TO AVOID LIMITING PRECISION DIFFICULTIES ARISING
C     FROM IMPOSSIBLE ACCURACY REQUESTS
C
      RER=2.0*EPS+REMIN
      IF (RELERR .GE. RER) GO TO 55
C
C     RELATIVE ERROR TOLERANCE TOO SMALL
      RELERR=RER
      IFLAG=3
      KFLAG=3
      RETURN
C
   55 DT=TOUT-T
C
      IF (MFLAG .EQ. 1) GO TO 60
      IF (INIT .EQ. 0) GO TO 65
      GO TO 80
```

```
C
C       INITIALIZATION --
C                       SET INITIALIZATION COMPLETION INDICATOR,INIT
C                       SET INDICATOR FOR TOO MANY OUTPUT POINTS,KOP
C                       EVALUATE INITIAL DERIVATIVES
C                       SET COUNTER FOR FUNCTION EVALUATIONS,NFE
C                       ESTIMATE STARTING STEPSIZE
C
   60 INIT=0
      KOP=0
C
      A=T
      CALL F(A,Y,YP)
      NFE=1
      IF (T .NE. TOUT) GO TO 65
      IFLAG=2
      RETURN
C
C
   65 INIT=1
      H=ABS(DT)
      TOLN=0.
      DO 70 K=1,NEQN
        TOL=RELERR*ABS(Y(K))+ABSERR
        IF (TOL .LE. 0.) GO TO 70
        TOLN=TOL
        YPK=ABS(YP(K))
        IF (YPK*H**5 .GT. TOL) H=(TOL/YPK)**0.2
   70 CONTINUE
      IF (TOLN .LE. 0.0) H=0.0
      H=AMAX1(H,U26*AMAX1(ABS(T),ABS(DT)))
      JFLAG=ISIGN(2,IFLAG)
C
C
C     SET STEPSIZE FOR INTEGRATION IN THE DIRECTION FROM T TO TOUT
C
   80 H=SIGN(H,DT)
C
C     TEST TO SEE IF RKF45 IS BEING SEVERELY IMPACTED BY TOO MANY
C     OUTPUT POINTS
C
      IF (ABS(H) .GE. 2.0*ABS(DT)) KOP=KOP+1
      IF (KOP .NE. 100) GO TO 85
C
C     UNNECESSARY FREQUENCY OF OUTPUT
      KOP=0
      IFLAG=7
      RETURN
C
```

```
   85 IF (ABS(DT) .GT. U26*ABS(T)) GO TO 95
C
C      IF TOO CLOSE TO OUTPUT POINT,EXTRAPOLATE AND RETURN
C
       DO 90 K=1,NEQN
   90  Y(K)=Y(K)+DT*YP(K)
       A=TOUT
       CALL F(A,Y,YP)
       NFE=NFE+1
       GO TO 300
C
C
C      INITIALIZE OUTPUT POINT INDICATOR
C
   95 OUTPUT= .FALSE.
C
C      TO AVOID PREMATURE UNDERFLOW IN THE ERROR TOLERANCE FUNCTION,
C      SCALE THE ERROR TOLERANCES
C
       SCALE=2.0/RELERR
       AE=SCALE*ABSERR
C
C
C      STEP BY STEP INTEGRATION
C
  100 HFAILD= .FALSE.
C
C      SET SMALLEST ALLOWABLE STEPSIZE
C
       HMIN=U26*ABS(T)
C
C      ADJUST STEPSIZE IF NECESSARY TO HIT THE OUTPUT POINT.
C      LOOK AHEAD TWO STEPS TO AVOID DRASTIC CHANGES IN THE STEPSIZE AND
C      THUS LESSEN THE IMPACT OF OUTPUT POINTS ON THE CODE.
C
       DT=TOUT-T
       IF (ABS(DT) .GE. 2.0*ABS(H)) GO TO 200
       IF (ABS(DT) .GT. ABS(H)) GO TO 150
C
C      THE NEXT SUCCESSFUL STEP WILL COMPLETE THE INTEGRATION TO THE
C      OUTPUT POINT
C
       OUTPUT= .TRUE.
       H=DT
       GO TO 200
C
  150 H=0.5*DT
C
C
```

```
C
C      CORE INTEGRATOR FOR TAKING A SINGLE STEP
C
C      THE TOLERANCES HAVE BEEN SCALED TO AVOID PREMATURE UNDERFLOW IN
C      COMPUTING THE ERROR TOLERANCE FUNCTION ET.
C      TO AVOID PROBLEMS WITH ZERO CROSSINGS,RELATIVE ERROR IS MEASURED
C      USING THE AVERAGE OF THE MAGNITUDES OF THE SOLUTION AT THE
C      BEGINNING AND END OF A STEP.
C      THE ERROR ESTIMATE FORMULA HAS BEEN GROUPED TO CONTROL LOSS OF
C      SIGNIFICANCE.
C      TO DISTINGUISH THE VARIOUS ARGUMENTS, H IS NOT PERMITTED
C      TO BECOME SMALLER THAN 26 UNITS OF ROUNDOFF IN T.
C      PRACTICAL LIMITS ON THE CHANGE IN THE STEPSIZE ARE ENFORCED TO
C      SMOOTH THE STEPSIZE SELECTION PROCESS AND TO AVOID EXCESSIVE
C      CHATTERING ON PROBLEMS HAVING DISCONTINUITIES.
C      TO PREVENT UNNECESSARY FAILURES, THE CODE USES 9/10 THE STEPSIZE
C      IT ESTIMATES WILL SUCCEED.
C      AFTER A STEP FAILURE, THE STEPSIZE IS NOT ALLOWED TO INCREASE FOR
C      THE NEXT ATTEMPTED STEP. THIS MAKES THE CODE MORE EFFICIENT ON
C      PROBLEMS HAVING DISCONTINUITIES AND MORE EFFECTIVE IN GENERAL
C      SINCE LOCAL EXTRAPOLATION IS BEING USED AND EXTRA CAUTION SEEMS
C      WARRANTED.
C
C
C      TEST NUMBER OF DERIVATIVE FUNCTION EVALUATIONS.
C      IF OKAY,TRY TO ADVANCE THE INTEGRATION FROM T TO T+H
C
  200 IF (NFE .LE. MAXNFE) GO TO 220
C
C      TOO MUCH WORK
       IFLAG=4
       KFLAG=4
       RETURN
C
C      ADVANCE AN APPROXIMATE SOLUTION OVER ONE STEP OF LENGTH H
C
  220 CALL  FEHL(F,NEQN,Y,T,H,YP,F1,F2,F3,F4,F5,F1)
       NFE=NFE+5
C
C      COMPUTE AND TEST ALLOWABLE TOLERANCES VERSUS LOCAL ERROR ESTIMATES
C      AND REMOVE SCALING OF TOLERANCES. NOTE THAT RELATIVE ERROR IS
C      MEASURED WITH RESPECT TO THE AVERAGE OF THE MAGNITUDES OF THE
C      SOLUTION AT THE BEGINNING AND END OF THE STEP.
C
       EEOET=0.0
       DO 250 K=1,NEQN
         ET=ABS(Y(K))+ABS(F1(K))+AE
         IF (ET .GT. 0.0) GO TO 240
C
C        INAPPROPRIATE ERROR TOLERANCE
         IFLAG=5
         RETURN
```

```
C
  240    EE=ABS((-2090.0*YP(K)+(21970.0*F3(K)-15048.0*F4(K)))+
     1            (22528.0*F2(K)-27360.0*F5(K)))
  250    EEOET=AMAX1(EEOET,EE/ET)
C
        ESTTOL=ABS(H)*EEOET*SCALE/752400.0
C
        IF (ESTTOL .LE. 1.0) GO TO 260
C
C
C       UNSUCCESSFUL STEP
C                      REDUCE THE STEPSIZE , TRY AGAIN
C                      THE DECREASE IS LIMITED TO A FACTOR OF 1/10
C
        HFAILD= .TRUE.
        OUTPUT= .FALSE.
        S=0.1
        IF (ESTTOL .LT. 59049.0) S=0.9/ESTTOL**0.2
        H=S*H
        IF (ABS(H) .GT. HMIN) GO TO 200
C
C       REQUESTED ERROR UNATTAINABLE AT SMALLEST ALLOWABLE STEPSIZE
        IFLAG=6
        KFLAG=6
        RETURN
C
C
C       SUCCESSFUL STEP
C                      STORE SOLUTION AT T+H
C                      AND EVALUATE DERIVATIVES THERE
C
  260 T=T+H
      DO 270 K=1,NEQN
  270   Y(K)=F1(K)
      A=T
      CALL F(A,Y,YP)
      NFE=NFE+1
C
C
C                      CHOOSE NEXT STEPSIZE
C                      THE INCREASE IS LIMITED TO A FACTOR OF 5
C                      IF STEP FAILURE HAS JUST OCCURRED, NEXT
C                          STEPSIZE IS NOT ALLOWED TO INCREASE
C
      S=5.0
      IF (ESTTOL .GT. 1.889568E-4) S=0.9/ESTTOL**0.2
      IF (HFAILD) S=AMIN1(S,1.0)
      H=SIGN(AMAX1(S*ABS(H),HMIN),H)
C
C     END OF CORE INTEGRATOR
C
```

```
C
C      SHOULD WE TAKE ANOTHER STEP
C
       IF (OUTPUT) GO TO 300
       IF (IFLAG .GT. 0) GO TO 100
C
C
C      INTEGRATION SUCCESSFULLY COMPLETED
C
C      ONE-STEP MODE
       IFLAG=-2
       RETURN
C
C      INTERVAL MODE
   300 T=TOUT
       IFLAG=2
       RETURN
C

       END
       SUBROUTINE  FEHL(F,NEQN,Y,T,H,YP,F1,F2,F3,F4,F5,S)
C
C      FEHLBERG FOURTH-FIFTH ORDER RUNGE-KUTTA METHOD
C
C      FEHL INTEGRATES A SYSTEM OF NEQN FIRST ORDER
C      ORDINARY DIFFERENTIAL EQUATIONS OF THE FORM
C             DY(I)/DT=F(T,Y(1),---,Y(NEQN))
C      WHERE THE INITIAL VALUES Y(I) AND THE INITIAL DERIVATIVES
C      YP(I) ARE SPECIFIED AT THE STARTING POINT T. FEHL ADVANCES
C      THE SOLUTION OVER THE FIXED STEP H AND RETURNS
C      THE FIFTH ORDER (SIXTH ORDER ACCURATE LOCALLY) SOLUTION
C      APPROXIMATION AT T+H IN ARRAY S(I).
C      F1,---,F5 ARE ARRAYS OF DIMENSION NEQN WHICH ARE NEEDED
C      FOR INTERNAL STORAGE.
C      THE FORMULAS HAVE BEEN GROUPED TO CONTROL LOSS OF SIGNIFICANCE.
C      FEHL SHOULD BE CALLED WITH AN H NOT SMALLER THAN 13 UNITS OF
C      ROUNDOFF IN T SO THAT THE VARIOUS INDEPENDENT ARGUMENTS CAN BE
C      DISTINGUISHED.
C
C
       INTEGER  NEQN
       REAL   Y(NEQN),T,H,YP(NEQN),F1(NEQN),F2(NEQN),
      1    F3(NEQN),F4(NEQN),F5(NEQN),S(NEQN)
C
       REAL   CH
       INTEGER   K
C
       CH=H/4.0
       DO 221 K=1,NEQN
   221     F5(K)=Y(K)+CH*YP(K)
       CALL F(T+CH,F5,F1)
C
```

```
      CH=3.0*H/32.0
      DO 222 K=1,NEQN
  222   F5(K)=Y(K)+CH*(YP(K)+3.0*F1(K))
      CALL F(T+3.0*H/8.0,F5,F2)
C
      CH=H/2197.0
      DO 223 K=1,NEQN
  223   F5(K)=Y(K)+CH*(1932.0*YP(K)+(7296.0*F2(K)-7200.0*F1(K)))
      CALL F(T+12.0*H/13.0,F5,F3)
C
      CH=H/4104.0
      DO 224 K=1,NEQN
  224   F5(K)=Y(K)+CH*((8341.0*YP(K)-845.0*F3(K))+
     1                      (29440.0*F2(K)-32832.0*F1(K)))
      CALL F(T+H,F5,F4)
C
      CH=H/20520.0
      DO 225 K=1,NEQN
  225   F1(K)=Y(K)+CH*((-6080.0*YP(K)+(9295.0*F3(K)-
     1         5643.0*F4(K)))+(41040.0*F1(K)-28352.0*F2(K)))
      CALL F(T+H/2.0,F1,F5)
C
C     COMPUTE APPROXIMATE SOLUTION AT T+H
C
      CH=H/7618050.0
      DO 230 K=1,NEQN
  230   S(K)=Y(K)+CH*((902880.0*YP(K)+(3855735.0*F3(K)-
     1         1371249.0*F4(K)))+(3953664.0*F2(K)+
     2         277020.0*F5(K)))
C
      RETURN
      END
```

PROBLEMS

P6-1. The error function is usually defined by an integral,

$$\operatorname{erf}(x) = \frac{2}{\sqrt{\pi}} \int_0^x e^{-t^2}\, dt,$$

but it can also be defined as the solution to the differential equation

$$y'(x) = \frac{2}{\sqrt{\pi}} e^{-x^2},$$

$$y(0) = 0.$$

(Why?) Write a program which uses RKF45 to print a table of erf (x) for $x = 0.0, 0.1, 0.2, \ldots, 1.9, 2.0$. Compare your table with the one produced for Problem P5-1. If you have access to precise execution times for programs run on your computer, measure the amount of time required by the two methods for generating the same table.

P6-2. A vehicle of mass M, suspended by a lumped spring-dashpot system as shown, travels with constant horizontal velocity. At time $t = 0$ the vehicle is traveling with its center of gravity at a distance h_0 above the ground and with no vertical velocity. For subsequent times, the vertical displacement of the road above reference level is given by a function $x_0(t)$.

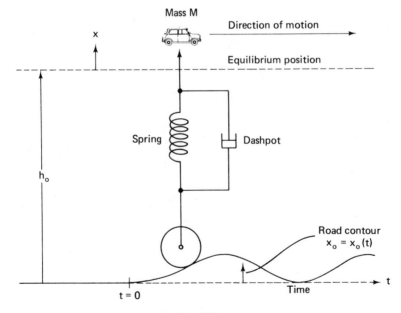

Fig. P6.2.

Suppose the spring is linear with a constant of proportionality k and that the damping coefficient of the dashpot r is a nonlinear function of the relative velocities of the two ends of the dashpot:

$$r = r_0\left(1 + c\left|\frac{dx}{dt} - \frac{dx_0}{dt}\right|\right).$$

It is easily shown that the displacement of the center of gravity of the vehicle $x(t)$ is the solution of the second-order ordinary differential equation

$$M\frac{d^2x}{dt^2} = -k(x - x_0) - r\left(\frac{dx}{dt} - \frac{dx_0}{dt}\right)$$

with the initial conditions

$$x(0) = 0,$$

$$\left.\frac{dx}{dt}\right|_{t=0} = 0.$$

Write a Fortran program to calculate $x(t)$, $0 \le t \le t_{max}$, for a road contour given by

$$x_0(t) = A(1 - \cos \omega t),$$

where $2A$ is the maximum road displacement above the reference level.

Note that for the linear case ($c = 0$) the underdamped, critically damped, and overdamped cases correspond to

$$\xi = \frac{r}{2\sqrt{kM}},$$

being less than, equal to, and greater than unity, respectively.

The program should read values for M, RO, C, K, A, W, TMAX, and FREQ, where W corresponds to ω. FREQ is used to control the printing frequency; i.e., the values of t, x_0, x, dx/dt, d^2x/dt^2, and $\xi(t)$ are printed every FREQ seconds.

Suggested test data:

$$M = 10 \text{ lb-sec}^2/\text{in}, \quad A = 2 \text{ in};$$

$$W = 7 \text{ rad/sec}, \quad \text{TMAX} = 5 \text{ sec}, \quad K = 640 \text{ lb}_f/\text{in}$$

Investigate values of

RO (lb$_f$-sec/in)	C (sec/in.)
80	0
160	1
240	10

P6-3. The fundamental equation of beam theory is

$$\frac{d^2y}{dx^2} = -\frac{M}{EI},$$

where x is the horizontal distance along the beam, y is the vertical downward deflection, M is the bending moment, E is Young's modulus, and I is the moment of area (sometimes called moment of inertia) of the cross section about the neutral axis. The moment of area I need not be constant, as the shape of the beam's cross section may change along the length (x dimension) of the beam.

It is easily shown that

$$\frac{dM}{dx} = V,$$

where V is the shearing force, and

$$\frac{dV}{dx} = -w,$$

where $w(x)$ is the load on the beam.

Thus, differentiating the fundamental equation twice and substituting, we obtain

$$\frac{d^4y}{dx^4} = \frac{-2}{I}\frac{dI}{dx}\frac{d^3y}{dx^3} - \frac{1}{I}\frac{d^2I}{dx^2}\frac{d^2y}{dx^2} + \frac{w}{EI}.$$

Most applications of this equation result in boundary-value problems. However, the following initial value problem is of practical interest.

Consider a cantilever beam of varying cross section, as shown. The length of the beam is l, and it supports a load P at its end. For the above beam,

$$V(x) = P$$

and

$$M(x) = -P(l - x).$$

Let

$$I(x) = 5(1 + 4e^{-6x/l}) \text{ in.}^4,$$
$$E = 30 \times 10^6 \text{ psi},$$
$$l = 100 \text{ in.},$$
$$P = 500 \text{ lb}_f.$$

Compute $y(l)$, assuming the beam does not break or permanently deform.

Hint: The "fixed" end of the cantilever beam at $x = 0$ implies $y(0) = 0$ and $y'(0) = 0$. The other two initial values at $x = 0$ can be found with the equations for $V(0)$ and $M(0)$.

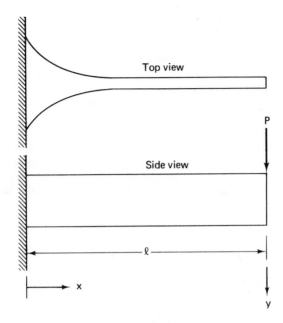

Fig. P6.3.

P6-4. Consider a simple ecosystem consisting of rabbits that have an infinite food supply and foxes that prey upon the rabbits for their food. A classical mathematical model due to Volterra describes this system by a pair of non-linear, first-order differential equations:

$$\frac{dr}{dt} = 2r - \alpha rf, \qquad r(0) = r_0,$$

$$\frac{df}{dt} = -f + \alpha rf, \qquad f(0) = f_0,$$

where t is time, $r = r(t)$ is the number of rabbits, $f = f(t)$ is the number of foxes, and α is a positive constant. When $\alpha = 0$, the two populations do not interact, and so the rabbits do what rabbits do best and the foxes die off from starvation. When $\alpha > 0$, the foxes encounter the rabbits with a probability which is proportional to the product of their numbers. Such an encounter results in a decrease in the number of rabbits and (for less obvious reasons) an increase in the number of foxes.

Investigate the behavior of this system for $\alpha = 0.01$ and various values of r_0 and f_0 ranging from 2 or 3 to several thousand. Draw graphs of interesting solutions, either by hand or with whatever computer plotting systems you have available. Include a plot with r as one axis and f as the other. (Since we are being rather vague about the units of measurement, there is no reason to restrict r and f to integer values.)

(a) Compute the solution with $r_0 = 300$ and $f_0 = 150$. You should observe from the output that the behavior of the system is periodic with a period very close to five time units. In other words, $r(5)$ is close to $r(0)$ and $f(5)$ is close to $f(0)$.

(b) Compute the solution with $r_0 = 15$ and $f_0 = 22$. You should find that the number of rabbits eventually drops below 1. This could be interpreted by saying that the rabbits become extinct. Find initial conditions which cause the foxes to become extinct. Find initial conditions with $r_0 = f_0$ which cause both species to become extinct.

(c) Is it possible for either component of the true solution to become negative? Is it possible for the numerical solution to become negative? What happens if it does? (In practice, the answers to the last two questions may depend on the values used for your error tolerances.)

(d) Many modifications of this simple model have been proposed to more accurately reflect what happens in nature. For example, the number of rabbits may be prevented from growing indefinitely by changing the first equation to

$$\frac{dr}{dt} = 2\left(1 - \frac{r}{R}\right)r - \alpha r f.$$

Now, even when $\alpha = 0$, the number of rabbits can never exceed R. Pick some reasonable value of R, and consider some of the above questions again. In particular, what happens to the periodicity of the solutions?

This model has been studied extensively by both mathematicians and biologists. The basic mathematical properties are described by Davis (1962). A comparison with actual observations of lynx and hare populations in the Hudson Bay area is made by Lotka (1956), pp. 88–94. Modifications of the model are discussed and further references given in articles by May (1972) and Gilpin (1972).

P6-5. A famous problem of nonlinear mechanics is known as the *inverted pendulum*. The pendulum is a stiff bar of length L which is supported at one end by a frictionless pin. The support pin is given a rapid up-and-down motion s by means of an electric motor:

$$s = A \sin \omega t.$$

A simple application of Newton's second law of motion yields the equation of motion

$$\ddot{\theta} = \frac{3}{2L}(g - A\omega^2 \sin \omega t) \sin \theta,$$

where g is the acceleration due to gravity. For small values of θ, $\sin \theta \approx \theta$, and this equation becomes the well-known Mathieu equation, which is known to be stable for certain values of A and ω and initial values. For $A = 0$ we have the familiar pendulum equation

$$\ddot{\theta} = \frac{3g}{2L} \sin \theta,$$

which can be linearized for values of θ near π.

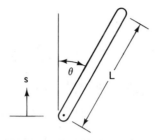

The most interesting aspect of this problem is that there are regions in which the equation of motion is stable for initial values corresponding to an inverted configuration, and they have been physically realized. Write a Fortran program using RKF45 to compute the motion $\theta(t)$ for various values of L, A, ω and initial values $\theta(0)$ and $\dot{\theta}(0)$. Debug your program using the case $L = 10$ in., $A = 0$ in., $\omega = 0$ rad/sec, $\theta(0) = 3.1$ rad, and $\dot{\theta}(0) = 0$ rad/sec, where g is taken to be 386.09 in./sec². Compare your computed solution with the analytical solution of the linearized version. You should use the analytical solution to determine good values for the various parameters required by RKF45. Print θ and $\dot{\theta}$ for two or three oscillations of the pendulum.

When your program appears to be working satisfactorily, try the more interesting cases:

L	A	ω	$\theta(0)$	$\dot{\theta}(0)$
10	0.5	5.3	3.1	0
10	10	100	3.1	0
10	10	100	0.1	0
10	2	100	0.1	0
10	0.5	200	0.05	0

Interpret your solutions physically.

P6-6. The following differential equations describe the motion of a body in orbit about two much heavier bodies. An example would be an Apollo capsule in an earth-moon orbit. The coordinate system is a little tricky. The three bodies determine a plane in space and a two-dimensional cartesian coordinate in this plane. The origin is at the center of mass of the two heavy bodies, the x axis is the line through these two bodies, and the distance between them is taken as the unit. Thus, if μ is the ratio of the mass of the moon to that of the earth, then the moon and the earth are located at coordinates $(1 - \mu, 0)$ and $(-\mu, 0)$, respectively, and the coordinate system moves as the moon rotates about the earth. The third body, the Apollo, is assumed to have a mass which is negligible compared to the other two, and its position as a function of time is $(x(t), y(t))$. The equations are derived from Newton's law of motion and the inverse square law of gravitation. The first-order derivatives in the equation come from the rotating coordinate system.

$$x'' = 2y' + x - \frac{\mu^*(x + \mu)}{r_1^3} - \frac{\mu(x - \mu^*)}{r_2^3},$$

$$y'' = -2x' + y - \frac{\mu^* y}{r_1^3} - \frac{\mu y}{r_2^3},$$

$$\mu = \frac{1}{82.45}, \qquad \mu^* = 1 - \mu,$$

$$r_1 = ((x + \mu)^2 + y^2)^{1/2}, \qquad r_2 = ((x - \mu^*)^2 + y^2)^{1/2}.$$

Although a great deal is known about these equations, it is not possible to find closed-form solutions. One interesting class of problems involves the study of periodic solutions. It is known that the initial conditions

$$x(0) = 1.2, \qquad x'(0) = 0,$$
$$y(0) = 0, \qquad y'(0) = -1.04935751$$

lead to a solution which is periodic with period $T = 6.19216933$. This means the Apollo starts on the far side of the moon with an altitude of about 0.2 times the earth-moon distance and a certain initial velocity. The resulting orbit brings the Apollo in fairly close to the earth, out in a big loop on the opposite side of the earth from the moon, back in close to the earth again, and finally back to its original position and velocity on the far side of the moon.

(a) Use RKF45 to compute the solution with the given initial conditions. Verify that the solution is periodic with the given period.

(b) How close does the Apollo come to the *surface* of the earth in this orbit? In the equation, distances are measured from the centers of the earth and moon. Assume that the moon is 238,000 miles from the earth and that the earth is a sphere with radius of 4000 miles. Note that the coordinate system origin is within this sphere but not at its center.

(c) Modify the subroutine RKFS used by RKF45 so that it prints the position of the Apollo at the beginning of each step. Compute one complete orbit, and observe the changes in step size. You should notice that fairly small steps are taken when the Apollo is near the earth and its orbit is changing rapidly and that fairly large steps are taken when the Apollo is deep in space. Try several different values of the error tolerances. Estimate how much more work would be needed to obtain the same accuracy with a method which uses fixed step size. If you have access to plotting equipment, produce something like Fig. 5.1 which shows the varying step size.

P6-7. Write a program which obtains the total derivatives of a function $f(y, t)$. These are defined by

$$f^{(0)} = f,$$

$$f^{(k+1)} = \frac{\partial f^{(k)}}{\partial y} \cdot f + \frac{\partial f^{(k)}}{\partial t}.$$

(a) Use an arithmetic language like Fortran or Algol. Assume $f(y, t)$ is a polynomial in two variables, represented as a two-dimensional array of its coefficients.

(b) Use a string manipulation language like Snobol. What class of functions $f(y, t)$ can be handled, and how may they be represented?

(c) If available, use a symbolic algebra system like Formac, Altran, Reduce, Scratchpad, or Macsyma.

7
SOLUTION OF
NONLINEAR EQUATIONS

Let f be a polynomial or transcendental function of one real or complex variable. The problem is to find one or more *zeros* of f, i.e., solutions of the equation $f(x) = 0$. Finding formulas for zeros of polynomials was one of the biggest parts of Renaissance mathematics in Italy. For polynomials of degree 2, 3, or 4, algorithms were found centuries ago for expressing the roots in terms of a finite number of square roots or cube roots, together with rational operations. It was not until the 1830s that Galois showed the impossibility of similar algorithms for polynomials of degree 5 or higher, even when nth roots are permitted in the formulas.

Of course, with a computer we hardly care whether the problem is solvable in the Renaissance sense. The algorithms used are iterative, and the important questions are how many evaluations of $f(x)$ are needed, whether evaluations of such derivatives as $f'(x)$ and $f''(x)$ are used, and so on.

In Sections 7.1 and 7.2 we shall discuss methods for getting real zeros of general functions f. In Section 7.3 we shall deal with getting complex zeros. Special methods for getting zeros of polynomials are briefly mentioned in Section 7.4.

A generalization of this problem involves finding some or all of the solutions of a system of n simultaneous nonlinear algebraic or transcendental equations in n unknowns. This problem is discussed briefly in Section 7.5.

7.1. TRANSCENDENTAL EQUATIONS—REAL ROOTS

Let $f(x)$ be a function of a single real variable x. To start a search for a zero of $f(x)$, we assume that an interval $[a, b]$ can be found on which $f(x)$ changes sign. However, if we know nothing else about f, then we do not even know that it has a zero in the interval. Even if the *mathematical* function is continuous and changes sign in [a, b], the *computed* floating-point-valued function of a floating-point variable assumes values in only a discrete set of points, and possibly none of these is zero. Hence, ultimately it is more practical to search not for a zero of f but for a *small* interval [α, β] in which f changes sign. Such an interval can always be found, and it can be made as small as the floating-point number system permits—i.e., with successive numbers as end points. If nothing is known about f, the *method of bisection* is the safest algorithm to use. This involves a given tolerance ϵ and the following steps:

1. Let $\alpha = a$ and $\beta = b$. Compute $f(\alpha)$ and $f(\beta)$.
2. Let $\gamma = (\alpha + \beta)/2$. Compute $f(\gamma)$.
3. If sign $(f(\gamma)) =$ sign $(f(\alpha))$, then replace α by γ, otherwise, replace β by γ.
4. If $\beta - \alpha > \epsilon$, then go to step 2; otherwise, stop.

Clearly one bit of precision is gained in each iteration of the bisection algorithm. Thus t iterations are needed for a criterion $\beta - \alpha \leq 2^{-t}(b - a)$. With long-precision arithmetic on the IBM 360/370 approximately 56 iterations would be required to bring an initial interval [1, 16] down to two successive floating-point numbers.

The bisection algorithm is somewhat slow but is absolutely sure not to fail. If each evaluation of $f(x)$ is rapid, there is usually no strong reason to reject the bisection method. The only reason we do not use it is that we have another algorithm, ZEROIN, that is guaranteed to never be much slower than bisection and is faster when f is a smooth function. The added speed is very useful when the evaluation of $f(x)$ is expensive in time.

When the mathematical function f is smooth enough to have one or two continuous derivatives, it is often possible to reduce the number of function evaluations a great deal over the method of bisection. A large number of different iterative methods have been studied, of which we shall here discuss only *Newton's method* and the *method of secants*. For further reading about these and other methods, see J. F. Traub's article, "The Solution of Transcendental Equations," in Ralston and Wilf (1967).

In the method of Newton (sometimes named for both Newton and Raphson), a zero r of f is found as the limit of a sequence of real numbers $\{x_k\}$. (We are assuming exact arithmetic with real numbers.) Each new iterate x_{k+1} is found as the unique zero of the tangent line to the curve $y = f(x)$ at x_k, i.e., the *local linearization* of f at x_k. As is shown in calculus,

$$x_{k+1} = x_k - \frac{f(x_k)}{f'(x_k)},$$

provided $f'(x_k) \neq 0$.

Theorem

   If $f(r) = 0$ but $f'(r) \neq 0$ and if f'' is continuous, then there is an open interval $N(r)$ containing r such that if x_1 is in $N(r)$, then for Newton's method $x_k \longrightarrow r$ as $k \longrightarrow \infty$. Moreover, if we denote the error of x_k by $e_k = x_k - r$, then

$$\lim_{k \to \infty} \frac{e_{k+1}}{e_k^2} = \frac{f''(r)}{2f'(r)}.$$

   Instead of proving the theorem, we shall discuss it. The hypothesis states that r is a simple zero of f. The conclusion states that e_{k+1} is roughly equal to Ce_k^2, where $C = f''(r)/2f'(r)$. So each iteration approximately squares the error e_k. As $e_k \longrightarrow 0$, the number of good decimal (or binary) digits is approximately doubled at each iteration. If C is about 1, then as long as x_k is close enough to r so that $|e_k| < \frac{1}{2}$, then

$$|e_{k+1}| < \frac{1}{2^2}, |e_{k+2}| < \frac{1}{2^4}, \ldots, |e_{k+6}| < \frac{1}{2^{64}}.$$

Thus, roughly speaking, it takes about six iterations to move the error from $\frac{1}{2}$ down to the smallest error possible in IBM 360 long-precision floating-point arithmetic. This contrasts with the approximately 56 iterations needed in the method of bisection to achieve the same accuracy.

   Any convergent iteration for which the error e_k satisfies

$$\lim_{k \to \infty} \frac{e_{k+1}}{e_k^d} = C \neq 0$$

is said to have *convergence of order d*. With the hypothesis of the theorem, if $f''(r) \neq 0$, Newton's method has convergence of order 2, sometimes also called *quadratic convergence*.

   As soon as the error in Newton's method gets down to the order of the distance between nearby floating-point numbers, the granular structure of the number system makes it no longer possible to continue the algorithm.

   The real difficulty with Newton's method is finding a first guess x_1 that is within the interval $N(r)$ of the desired zero r. If the graph of f looks as shown in Fig. 7.1, then $N(r)$ is approximately the interval $(r - \epsilon, r + \epsilon)$. If x_1 is outside $N(r)$, the successive iterates in Newton's method go farther and farther away from r, and the zero is never found.

   Hence, frequently the Newton method must be preceded by some *globally convergent* algorithm like bisection before the rapidly convergent Newton method can take hold. So Newton's method may often in fact be only a termination procedure for a slower but surer initial algorithm. With its use, perhaps the last 25 or so iterative steps of bisection, for example, could be replaced by 6 Newton steps.

   If r is not a simple root, so that $f'(r) = 0$, the hypothesis of the theorem is

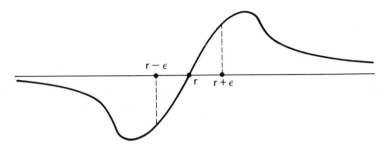

Fig. 7.1.

violated. The convergence of Newton's method to a double root is of order 1 (called *linear convergence*), not 2. For example, if $f(x) = x^2$, then in the Newton process $x_{k+1} = x_k/2$, whence $e_{k+1}/e_k = \frac{1}{2}$, for all k.

Note that each iteration of Newton's method requires the evaluation not only of $f(x)$ but also of $f'(x)$. There are functions for which it is very cheap to evaluate $f'(x)$ once $f(x)$ has been evaluated. For other functions, evaluating $f'(x)$ is as expensive as a second evaluation of $f(x)$. For still other functions, evaluating $f'(x)$ is almost impossible.

The main importance of Newton's method lies in the facts that it can be used to find complex zeros of analytic functions f and also that it can be extended to the solution of a simultaneous set of nonlinear equations in many variables.

When seeking the zeros of a function f for which the evaluation of $f'(x)$ is difficult, the *secant method* is often a better choice than Newton's method. In this algorithm one starts with two initial numbers x_1 and x_2. At each stage x_{k+1} is found from x_k and x_{k-1} as the unique zero of the linear function that has the value $f(x_k)$ at x_k and the value $f(x_{k-1})$ at x_{k-1}. This linear function represents the line that is the secant to the curve $y = f(x)$ through the points with abscissas x_k and x_{k-1}—hence the name *secant method*. See Fig. 7.2.

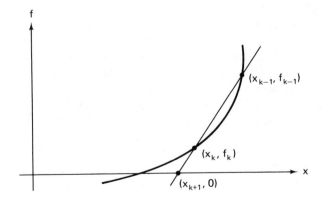

Fig. 7.2.

It is easily shown that

$$x_{k+1} = x_k - \frac{f_k}{(f_{k-1} - f_k)/(x_{k-1} - x_k)},$$

where $f_k = f(x_k)$. (It is better not to combine the term x_k with the next term. Why?) The convergence theorem for the method of secants is as follows:

Theorem

If $f(r) = 0$ but $f'(r) \neq 0$ and $f''(r) \neq 0$ and if f'' is continuous, then there is an open interval $N(r)$ containing r such that if x_1 and x_2 are distinct points in $N(r)$, then the sequence x_k converges to r, as $k \longrightarrow \infty$. Moreover,

$$\lim_{k \to \infty} \frac{e_{k+1}}{e_k^p} = C \neq 0,$$

where $p = \frac{1}{2}(\sqrt{5} + 1) \approx 1.618$.

The proof is rather more difficult than the easy proof possible for Newton's method. The conclusion states that for a good enough start, the secant method converges to a simple zero of a function with a continuous second derivative and that the convergence is of order 1.618.

In Newton's method, if you consider that two function evaluations (one for f and one for f') are needed at each step, then the order of convergence may be thought of as $\sqrt{2} \approx 1.414$ *per function evaluation*. Since only one function evaluation is needed per step of the secant method, the secant method can be considered to have better convergence than Newton's method.

Like Newton's method, the secant method works perfectly well for analytic functions of a complex variable. However, extending the secant method to systems of equations, while possible, seems quite difficult.

As with Newton's method, the biggest problem with the secant method is to find x_1 and x_2 close enough to r so that convergence can start.

When $f(x)$ has been evaluated at more than two points, it seems reasonable to use that information to improve later estimates of the zero. One such method is *inverse quadratic interpolation*, which uses three points, x_{k-2}, x_{k-1}, and x_k. Let $g(y)$ be the quadratic in a variable y for which $x_i = g(f_i)$, $i = k - 2, k - 1, k$. Then the next approximate zero is taken to be $x_{k+1} = g(0)$. This can be expressed directly in terms of the three x values and the three f values, but the actual formula is not important here. It is necessary for the three f values to be distinct. If they are not, a division by zero occurs in the formula.

The rate of convergence of inverse quadratic interpolation is 1.839, which is slightly faster than the secant method. However, three starting values are required, and the behavior of the algorithm can be very erratic when the starting values are not close enough to a zero.

7.2. SUBROUTINE ZEROIN

One of the best computer algorithms that we have for finding a real zero of a single function combines the certainty of bisection with the ultimate speed of the secant and inverse quadratic interpolation methods for smooth functions. It is called ZEROIN and was originated by van Wijngaarden, Dekker, and others at the Mathematical Center in Amsterdam in the 1960s. A description and analysis were given by Wilkinson (1976), especially pp. 8–12. The algorithm was first published by Dekker (1969) and improved by Brent (1973).

We shall use a Fortran implementation of Brent's version of the Dekker algorithm. A copy of the function subprogram ZEROIN appears at the end of this section. A typical call of ZEROIN might look like

$$ZZ = ZEROIN\ (A, B, F, TOL).$$

Here A, B are end points of the interval in which the zero is sought. The parameter F is a real function subprogram having one real variable as an argument. TOL is the amount of error which can be tolerated in the result. The program assumes without a test that F(A) and F(B) have different signs.

ZEROIN carries out an iteration in which three abscissas A, B, and C are present at each stage. Normally,

1. B is the latest iterate and closest approximation to the zero.
2. A is the previous iterate.
3. C is the previous or an older iterate so that F(B) and F(C) have opposite signs.

At all times B and C bracket the zero. Moreover $|F(B)| \leq |F(C)|$. When the interval length $|B - C|$ has been reduced to satisfy

$$|B - C| \leq TOL + 4.*EPS*ABS(B)$$

the value of B is returned as the value of ZEROIN. In addition to TOL, the convergence test involves the machine accuracy parameter, EPS, to guard against the possibility that TOL is too small. In particular, TOL can be set to zero to force ZEROIN to find the smallest possible interval.

At each step, ZEROIN chooses the next iterate from two candidates— one obtained by the bisection algorithm and one obtained by an interpolation algorithm. Inverse quadratic interpolation is used whenever A, B, and C are distinct, and linear interpolation (the secant method) is used whenever they are not. If the point obtained by interpolation is "reasonable," it is chosen; otherwise the bisection point is chosen. The definition of "reasonable" is rather technical, but essentially it means that the point is inside the current interval and not too close to the end points. Consequently the length of the interval is guaranteed to decrease at each step and to decrease by a large factor when the function is well behaved. For complete details, see Brent (1973).

A few programming details in ZEROIN should be mentioned because they are applicable and important in other situations. The bisection point is computed by

$$XM = B + 0.5*(C - B)$$

rather than the conventional

$$XM = 0.5*(C + B).$$

To see why, take $C = 0.982$ and $B = 0.987$. Assume the calculations are being done on a machine with three decimal digit chopped floating-point arithmetic. Then $C + B$ is computed as 1.96, and the conventional formula gives a midpoint 0.980 which is outside the interval. The general principle is that it is best to arrange formulas so that they express the desired quantity as a small correction to a good approximation.

Careful attention is paid to underflow and overflow problems. For example, the test to ensure that F(B) and F(C) have different signs is *not* the conventional

$$F(B)*F(C) < 0.0.$$

If $F(B) = 10^{-40}$ and $F(C) = -10^{-40}$, then they have opposite signs, but on many computers the product will underflow and be set to zero, and the test will fail. The point obtained by the interpolation algorithms is expressed in the form

$$B + P/Q,$$

but the division is not done unless it is both necessary and safe to do so. When a bisection is to be done, then this quantity is not needed anyway.

In summary, what does Brent claim for his version of ZEROIN? First, it will always converge, even with floating-point arithmetic. Second, the number of function evaluations cannot exceed a number roughly equal to

$$\left[\log_2 \left(\frac{B - A}{TOL1}\right)\right]^2,$$

where TOL1 = 0.5*TOL + 2.0*EPS*ABS(B). Third, the zero R returned by the algorithm is such that F is guaranteed to change sign in a stated interval that is approximately [R − 2·TOL1, R + 2·TOL1]. Fourth, Brent tested the algorithm very widely on a great variety of functions and found that roughly 10 function evaluations were typically needed for smooth functions. Fifth, he never found a function requiring more than three times the number of evaluations needed for bisection—that is, greater than 170 on the IBM 360 for a root not near 0. Sixth, if F is smooth enough to have a continuous second derivative near a simple zero R of F, then the algorithm ZEROIN (if started near enough to R) will eventually stop doing bisections and home in on R by a process that is about like the secant process, with a degree of convergence at least 1.618. For proofs of these claims, see Brent (1973).

The following program illustrates the use of ZEROIN on a simple example, $f(x) = x^3 − 2x − 5$. It is easy to see from a rough graph of $f(x)$ that there is only one real zero and that it lies between 2 and 3.

The output is Z = 2.0945514815.

This example is a polynomial which could be treated by the methods of Section 7.4, but we have used it here because of its historical interest. The following excerpt from a letter written by de Morgan to Whewell in 1861 is contained in a 1924 book, *The Calculus of Observations*, by Whittaker and Robinson: "The reason I call $x^3 − 2x − 5 = 0$ a celebrated equation is because it was the one on which Wallis chanced to exhibit Newton's method when he first published it, in consequence of which *every* numerical solver has felt bound in duty to make it one of his examples. Invent a numerical method, neglect to show how it works on this equation, and you are a pilgrim who does not come in at the little wicket (*vide* J. Bunyan)."

```
C  SAMPLE PROGRAM FOR ZEROIN
C
       REAL FUNCTION F(X)
       REAL X
       F = X*(X*X - 2.) - 5.
       RETURN
       END
C
       EXTERNAL F
       REAL A,B,Z,TOL,ZEROIN
       A = 2.
       B = 3.
       TOL = 1.0E-10
       Z = ZEROIN(A,B,F,TOL)
       WRITE(6,1) Z
   1   FORMAT(3H Z=,F15.10)
       STOP
       END
```

```
      REAL FUNCTION ZEROIN(AX,BX,F,TOL)
      REAL AX,BX,F,TOL
C
C     A ZERO OF THE FUNCTION  F(X)  IS COMPUTED IN THE INTERVAL AX,BX
C
C INPUT..
C
C AX     LEFT ENDPOINT OF INITIAL INTERVAL
C BX     RIGHT ENDPOINT OF INITIAL INTERVAL
C F      FUNCTION SUBPROGRAM WHICH EVALUATES F(X) FOR ANY X IN
C        THE INTERVAL AX,BX
C TOL    DESIRED LENGTH OF THE INTERVAL OF UNCERTAINTY OF THE
C        FINAL RESULT ( .GE. 0.0)
C
C
C OUTPUT..
C
C ZEROIN ABCISSA APPROXIMATING A ZERO OF  F  IN THE INTERVAL AX,BX
C
C
C     IT IS ASSUMED THAT  F(AX)   AND   F(BX)   HAVE  OPPOSITE  SIGNS
C WITHOUT  A  CHECK.  ZEROIN  RETURNS A ZERO  X  IN THE GIVEN INTERVAL
C AX,BX  TO WITHIN A TOLERANCE  4*MACHEPS*ABS(X) + TOL, WHERE MACHEPS
C IS THE RELATIVE MACHINE PRECISION.
C     THIS FUNCTION SUBPROGRAM IS A SLIGHTLY  MODIFIED  TRANSLATION  OF
C THE ALGOL 60 PROCEDURE  ZERO GIVEN IN  RICHARD BRENT, ALGORITHMS FOR
C MINIMIZATION WITHOUT DERIVATIVES, PRENTICE - HALL, INC. (1973).
C
C
      REAL   A,B,C,D,E,EPS,FA,FB,FC,TOL1,XM,P,Q,R,S
C
C COMPUTE EPS, THE RELATIVE MACHINE PRECISION
C
      EPS = 1.0
   10 EPS = EPS/2.0
      TOL1 = 1.0 + EPS
      IF (TOL1 .GT. 1.0) GO TO 10
C
C INITIALIZATION
C
      A = AX
      B = BX
      FA = F(A)
      FB = F(B)
```

```
C
C BEGIN STEP
C
   20 C = A
      FC = FA
      D = B - A
      E = D
   30 IF (ABS(FC) .GE. ABS(FB)) GO TO 40
      A = B
      B = C
      C = A
      FA = FB
      FB = FC
      FC = FA
C
C CONVERGENCE TEST
C
   40 TOL1 = 2.0*EPS*ABS(B) + 0.5*TOL
      XM = .5*(C - B)
      IF (ABS(XM) .LE. TOL1) GO TO 90
      IF (FB .EQ. 0.0) GO TO 90
C
C IS BISECTION NECESSARY
C
      IF (ABS(E) .LT. TOL1) GO TO 70
      IF (ABS(FA) .LE. ABS(FB)) GO TO 70
C
C IS QUADRATIC INTERPOLATION POSSIBLE
C
      IF (A .NE. C) GO TO 50
C
C LINEAR INTERPOLATION
C
      S = FB/FA
      P = 2.0*XM*S
      Q = 1.0 - S
      GO TO 60
C
C INVERSE QUADRATIC INTERPOLATION
C
   50 Q = FA/FC
      R = FB/FC
      S = FB/FA
      P = S*(2.0*XM*Q*(Q - R) - (B - A)*(R - 1.0))
      Q = (Q - 1.0)*(R - 1.0)*(S - 1.0)
```

```
C
C ADJUST SIGNS
C
   60 IF (P .GT. 0.0) Q = -Q
      P = ABS(P)
C
C IS INTERPOLATION ACCEPTABLE
C
      IF ((2.0*P) .GE. (3.0*XM*Q - ABS(TOL1*Q))) GO TO 70
      IF (P .GE. ABS(0.5*E*Q)) GO TO 70
      E = D
      D = P/Q
      GO TO 80
C
C BISECTION
C
   70 D = XM
      E = D
C
C COMPLETE STEP
C
   80 A = B
      FA = FB
      IF (ABS(D) .GT. TOL1) B = B + D
      IF (ABS(D) .LE. TOL1) B = B + SIGN(TOL1, XM)
      FB = F(B)
      IF ((FB*(FC/ABS(FC))) .GT. 0.0) GO TO 20
      GO TO 30
C
C DONE
C
   90 ZEROIN = B
      RETURN
      END
```

7.3. TRANSCENDENTAL EQUATIONS—COMPLEX ROOTS

Usually analytic functions will have not only real zeros but also complex zeros. Can the methods we have discussed in Sections 7.1 and 7.2 be used to find complex zeros?

The bisection method is based on the idea that a continuous function that changes sign in an interval will have a zero within the interval. It is not easy to generalize this idea for locating complex zeros of analytic functions. One related idea is the *principle of the argument*: Suppose f is analytic in a region R of the complex plane, and let C be a simple closed curve in R. Suppose, as z makes a circuit around C, that $f(z)$ loops the origin exactly once. Then f has exactly one zero inside C.

There have been various attempts to use this principle directly to locate zeros of f within a region, but none seems very successful. The basic trouble is the large number of times required to evaluate $f(z)$—to see whether the curve $\{f(z): z \in C\}$ loops the origin. D. H. Lehmer has used a sophisticated variant of the idea to locate zeros of polynomials. But for general functions f we do not seem to have any slow but sure analog of the method of bisection. Thus in practice people use other methods.

The Newton and secant methods can be used in the complex plane without any change in the theorems. As with real roots, it is necessary to get close to a zero before the rapid convergence can start. Frequently the originator of a problem knows from its application approximately where the zeros are, and this information can be used to start the iteration. Lacking any such guess, one often applies Newton's method from some random complex starting value z_0. In principle, three things can then happen to the sequence $\{z_k\}$:

1. It diverges to ∞.
2. It converges to a zero r.
3. It does neither, but just wanders around.

For certain functions in case 3 the $\{z_k\}$ will ultimately tend to a periodic state, in which the points nearly *cycle* with a period of 2 or 3 or some other finite number.

If you want a zero r of f, you can try applying Newton's method, with some device included to keep z_k from going too far away from z_{k-1}, and thus attempting to prevent divergence to ∞. Then you just hope that the method will converge to a zero. If, after some suitably large number of iterations, there has been no convergence, you can start over with some different z_0. Experience shows that often a zero will be found.

Once one zero r_1 has been found, if you want another, you need a way to prevent a continued iteration from returning to r_1. It is often useful to continue the iteration with the new function

$$f_1(z) = \frac{f(z)}{z - r_1},$$

where the division is made only between the *numbers* $f(z)$ and $z - r_1$. The key term in Newton's process is $f(z)/f'(z)$, whose reciprocal $f'(z)/f(z) = (d/dz) \ln f(z)$. Hence,

$$\frac{f_1'(z)}{f_1(z)} = \frac{d}{dz} \ln f(z) - \frac{d}{dz} \ln (z - r_1)$$

$$= \frac{f'(z)}{f(z)} - \frac{1}{z - r_1}.$$

After s roots, have been found, one applies the Newton algorithm to the function $f(z)/\prod_{k=1}^{s} (z - r_k)$, and logarithmic differentiation is still easy in principle.

If it is impossible to make any headway by direct application of the Newton or secant algorithms, one may have to precede them by a method of getting close to a zero. One idea is to use an algorithm for minimizing a real-valued function of two real variables x, y by applying it to the function

$$\phi(x, y) = |f(x + iy)|^2.$$

It is easy to prove that all local minima of $\phi(x, y)$ correspond to $\phi = 0$ if f is analytic. Methods of minimization are discussed in Chapter 8.

One much-used method for finding zeros of analytic function of a complex variable is due to Muller. This is an extension of the secant method from two points of interpolation to three points. Muller starts with arbitrary z_1, z_2, and z_3. At the general stage of his iteration there are three points z_{n-2}, z_{n-1}, and z_n and corresponding function values that we call f_{n-2}, f_{n-1}, and f_n. Muller forms the unique quadratic polynomial function of z that interpolates the three points (z_i, f_i), for $i = n - 2, n - 1$, and n. He then lets z_{n+1} be that one of the two zeros of the quadratic function that is nearer to z_n. The algorithm then proceeds with z_{n-1}, z_n, and z_{n+1}.

Muller's original motivation for the method was that the use of a quadratic would permit a zero finder starting with real iterates to proceed to complex iterates, in contrast to the secant or Newton methods. He then discovered that the method also worked well for finding real roots.

7.4. ZEROS OF POLYNOMIALS

Because polynomials are very special functions, it is understandable that many algorithms have been devised for finding their zeros. Indeed, methods for finding zeros of polynomials have constituted some of the oldest algorithms of numerical analysis. The methods of Horner, Graeffe, and Bernoulli date from former centuries, while in the computer era those of Rutishauser, Lehmer, Lin, Bairstow, Bareiss, and many others are known.

From the mathematical side there is much theoretical interest in producing algorithms that within the fictitious framework of real arithmetic can be proved never to fail to converge to a zero. From the practical side there is a demand for algorithms that will almost always work and will produce each zero in a small fraction of a second for polynomials of moderate degree on a large computer.

Compared with transcendental functions, polynomials have the advantage that one knows in advance exactly how many zeros there are and hence when to stop an algorithm.

The reader should recall the extreme sensitivity of the roots of some polynomials as functions of their coefficients: See Section 2.5. This means that many problems that involve finding the zeros of polynomials will require either extremely high-precision computation of the coefficients or else a totally different approach to the problem. For example, 20 or 30 years ago it was assumed that one would normally get the eigenvalues of a matrix by forming and then solving its characteristic equation. We now know much better ways to get the eigenvalues—ways that do not require high-precision computation (see Chapter 9). It is probable that many other problems where polynomials are used should be solved in other ways. So it may be that finding zeros of polynomials is no longer such an important part of scientific computing as it used to be.

The authors have had little experience with Fortran programs for this problem, but we suspect that Muller's method is quite good enough for many applications. A fast reliable Fortran program for finding zeros of polynomials with complex coefficients is given by Jenkins and Traub (1972).

7.5. NONLINEAR SYSTEMS OF EQUATIONS

A very frequent computational problem is to find some or all of the solutions of a system of n simultaneous nonlinear algebraic or transcendental equations in n unknowns. Letting $\mathbf{x}$ stand for the column vector $(x_1, \ldots, x_n)^T$, we can write the equations in the form $f_1(\mathbf{x}) = 0, f_2(\mathbf{x}) = 0, \ldots, f_n(\mathbf{x}) = 0$. Or, if we let $\mathbf{f}$ stand for a column vector of functions $(f_1, \ldots, f_n)^T$, then the entire system can be written in the condensed form $\mathbf{f}(\mathbf{x}) = \mathbf{0}$. Such a set of equations may arise directly in such a problem as designing a nonlinear physical system, or it may arise indirectly. For example, in trying to minimize a function $G(\mathbf{x})$, one may try to solve for the points where the gradient of G, and G, is zero. Letting $\mathbf{f} = \text{grad } G$, we have a nonlinear system.

One approach to solving the system $f(x) = 0$ is to generalize to n dimensions one of the iterative processes useful for solving a single equation (the case $n = 1$). If the partial derivatives of the functions f_i with respect to the variables x_j can all be computed, then Newton's process can be used. Let $J(x)$ stand for the Jacobian matrix, whose (i, j)th element is the value of $\partial f_i / \partial x_j$ at x. As in one dimension, the Newton idea is to start with an arbitrary x, say x^0. Then one linearizes the function f at x^0 by expanding f into a Taylor series and keeping only the terms of degrees 0 and 1:

$$f(x) = f(x^0) + J(x^0)(x - x^0) + \dots .$$

The linear approximation to f at x^0 is therefore given by

$$L(x) = f(x^0) + J^0(x - x^0),$$

where J^0 is an abbreviation for $J(x^0)$.

To find the next approximation x^1 to the solution of $f(x) = 0$, we solve the equation

$$f(x^0) + J^0(x^1 - x^0) = 0.$$

Of course the solution can be written in the form

$$x^1 = x^0 - (J^0)^{-1}f(x^0),$$

in which we see how closely the formula resembles the one-dimensional form of Newton's method. However, as with most systems of n linear equations in n unknowns, it is neither necessary nor even desirable to compute the inverse matrix $(J^0)^{-1}$; it is preferable just to solve the linear system for the correction $x^1 - x^0$.

In the general step of the iteration, having x^k, one obtains x^{k+1} by adding to x^k the correction $x^{k+1} - x^k$ obtained by solving the linear system

$$J^k(x^{k+1} - x^k) = -f(x^k).$$

A theorem analogous to the first one of Section 7.1 can be proved for this iteration. Speaking loosely, suppose r is a solution of $f(x) = 0$ at which $J(r)$ is not singular and that the second partials of f with respect to its arguments are continuous near r. Then if x^0 is sufficiently close to r, the Newton iteration will converge. Moreover, if $e^k = x^k - r$, then $\|e^{k+1}\|/\|e^k\|^2$ is bounded, as $k \to \infty$. Thus the convergence will be of order 2.

As in one dimension, the main problem is to get close enough to a desired root r to start the high-speed convergence. In practice, one can sometimes proceed with courage and optimism and find a root with little difficulty.

Another important consideration is the difficulty and expense of finding the Jacobian matrix. In one dimension, it is not unreasonable to assume that the cost of computing $f'(x)$ is about the same as that of computing $f(x)$, although there are certainly problems for which this is not true. In n dimensions, $\mathbf{f}(\mathbf{x})$ becomes a vector, while $\mathbf{f}'(\mathbf{x})$ becomes a matrix, the Jacobian. Hence it is often true that evaluation of $\mathbf{f}'(\mathbf{x})$ is many times more expensive than evaluation of $\mathbf{f}(\mathbf{x})$.

Attempts to avoid explicit use of the Jacobian have resulted in many different kinds of methods. Direct generalizations of the secant method are unsatisfactory because the approximations to $J(\mathbf{x})$ obtained from n-dimensional analogs of secants are often singular. See Gragg and Stewart (1976). More satisfactory methods are known as *quasi-Newton* methods. They generate approximations to $J(\mathbf{x})$ which are fairly crude at the beginning of the iterations but which are increasingly accurate as the iterations proceed. They are usually described as they apply to multidimensional optimization problems, which we shall consider in Section 8.3.

There is some discussion of the Newton process in Traub's article (cited earlier) in Ralston and Wilf (1967). A careful statement and proof of the convergence theorem is given by Ostrowski (1966). The book by Ortega and Rheinboldt (1970) is a complete reference for many of the theoretical results.

PROBLEMS

P7-1. Use ZEROIN to find x so that

$$\text{erf}(x) - 0.5 = 0.0.$$

Use any available subroutine to compute erf (x), or see Problem P5-1.

P7-2. Use ZEROIN to find the ten smallest positive values of x for which the line $y = x$ crosses the graph of $y = \tan x$.

P7-3. Let $f(x) = x(x - 1)^5$, and let A $= -0.50$ and B $= 0.98$. Apply ZEROIN and print out the values of A, B, and C for each iteration. Also print out an indication of whether the step was made by bisection, secant, or inverse quadratic interpolation.

P7-4. Use Newton's method to find a complex root of our historical example,

$$x^3 - 2x - 5 = 0.$$

P7-5. One classical method for solving cubics is Cardan's solution. The cubic equation

$$x^3 + ax^2 + bx + c = 0$$

is transformed to a reduced form

$$y^3 + py + q = 0$$

by the substitution $x = y - a/3$. The coefficients in the reduced form are

$$p = b - \frac{a^2}{3},$$

$$q = c - \frac{ab}{3} + 2\left(\frac{a}{3}\right)^3.$$

One real root of the reduced form can be found by

$$s = \left[\left(\frac{p}{3}\right)^3 + \left(\frac{q}{2}\right)^2\right]^{1/2},$$

$$y_1 = \left[-\frac{q}{2} + s\right]^{1/3} + \left[-\frac{q}{2} - s\right]^{1/3}$$

and then a real root of the original equation by

$$x_1 = y_1 - \frac{a}{3}.$$

The other two roots can be found by similar formulas or by dividing out x_1 and solving the resulting quadratic.

(a) Apply Cardan's method to find the real root of

$$x^3 + 3x^2 + \alpha^2 x + 3\alpha^2 = 0$$

for various values of α. Investigate the loss of accuracy from roundoff for large α, say α about the reciprocal of machine epsilon.

(b) Apply Newton's method to the same equation for the same values of α. Investigate the effects of roundoff error and the choice of starting value.

P7-6. What output is produced by the following program? Explain why.

```
        FUNCTION F(X)
        IF (X .EQ. O.) Y = -1.
        IF (X .GT. O.) Y = 9.
        WRITE(6,1) X,Y
   1    FORMAT(F15.10, F5.0)
        F = Y
        RETURN
        END

        EXTERNAL F
        A = 0.
        B = 1.
        TOL = 1.E-6
        Z = ZEROIN(A, B, F, TOL)
        STOP
        END
```

P7-7. This problem deals with the propagation of waves in a medium with variable speed of propagation. The particular setting is underwater propagation of sound, but the techniques are applicable to other situations. It involves a combination of data fitting, ordinary differential equation solving, and zero finding. Subroutines SPLINE, RKF45, and ZEROIN are used. The problem is derived from a paper by Moler and Solomon (1970).

The speed of sound in ocean water depends on pressure, temperature, and salinity, all of which vary with depth in fairly complicated ways. Let z denote depth in feet under the ocean surface (so that the positive z axis points down) and let $c(z)$ denote the speed of sound at depth z. We shall ignore the changes in sound speed observed in horizontal directions. It is possible to measure $c(z)$ at discrete values of z, and the following table is typical of those obtained:

z (ft)	$c(z)$ (ft/sec)
0	5,042
500	4,995
1,000	4,948
1,500	4,887
2,000	4,868
2,500	4,863
3,000	4,865
3,500	4,869
4,000	4,875
5,000	4,875
6,000	4,887
7,000	4,905
8,000	4,918
9,000	4,933
10,000	4,949
11,000	4,973
12,000	4,991

To obtain values of $c(z)$ between data points, we can use cubic spline interpolation. We shall also need to evaluate $c'(z) = dc(z)/dz$, so the first step of this problem is the following:

(a) Use SPLINE to obtain the coefficients of the cubic spline interpolating the above data. Modify SEVAL so that it returns both the value of the spline and the value of its derivative at any point.

Since the sound speed varies with depth, sound rays will travel in curved paths. The effect is a continuous version of the familiar refraction of light waves caused by the air-water interface in a fish bowl. The basic equation is a continuous version of Snell's law. Let x denote the horizontal (radial) distance in feet from a source of sound and $z(x)$ denote the depth of a particular ray at distance x. Let $\theta = \theta(x)$ denote the angle between a horizontal line and the tangent to the ray at x, i.e.,

$$\tan \theta = \frac{dz}{dx}.$$

Snell's law can be written

$$\frac{\cos \theta}{c(z)} = \text{constant}.$$

These two equations together yield a first-order ordinary differential equation which appears to define z as a function of x. However, it turns out that this equation fails to have unique solutions at those points where the ray becomes horizontal. To eliminate the unwanted solutions, we differentiate both equations with respect to x and combine the results to obtain a second-order equation

$$\frac{d^2z}{dx^2} = -\frac{c'(z)}{A^2 c(z)^3},$$

where A is the constant occurring in Snell's law. The constant and the initial conditions can be conveniently expressed in terms of z_0, the depth of the source, and θ_0, the ray angle at the source. We find

$$z(0) = z_0,$$

$$\frac{dz}{dx}(0) = \tan \theta_0,$$

$$A^2 = \left(\frac{\cos \theta_0}{c(z_0)}\right)^2.$$

This brings us to the second part of the problem.

(b) Use RKF45 to find the ray with $z_0 = 2000$ feet and $\theta_0 = 5.4$ degrees. Trace the ray over a distance of 24 nautical miles, printing its depth at intervals of 1 mile. Assume 1 nautical mile is 6076 feet and do not forget that the Fortran trigonometric functions expect arguments in radians. You should find that the depth at 24 miles is close to 3000 feet.

Now suppose that a sound source at a depth of 2000 feet transmits to a receiver 24 miles away at a depth of 3000 feet. The above calculation shows that one of the rays from the source to the receiver leaves the source at an angle close to 5.4 degrees. How many other rays are there? Let $x_f = 24$ nautical miles. As θ_0 varies, $z(x_f)$ also varies. We are interested in finding values of θ_0 for which $z(x_f) = 3000$.

(c) Write a function subprogram, F(THETA), which traces the ray with initial angle THETA and returns the value $z(x_f) - 3000$. Print a table of values of this function for THETA in the range -10 to 10 degrees. Since this function is fairly expensive to evaluate, the increment used in this table will depend on the amount of computer time you have available and the efficiency of your program.

(d) Use ZEROIN with starting values obtained from part (c) to find rays which pass through (or as near as possible to) the receiver.

(e) Assume that the floor of the ocean is at 12,000 feet and that the surface is at 0 feet. What happens if θ_0 is greater than 10 degrees? What if it is less than -10 degrees? Are there any rays with initial angles in these ranges?

(f) Discuss the effect of your choices of error tolerances in RKF45 and ZEROIN on the accuracy and the cost of your solutions. Consider using different tolerances in parts (c) and (d).

P7-8.   Suppose that you have an "uncovered" wagon as shown in the accompanying figure and that you must make a cover for it. The problem is to choose a length of thin wooden strips (ribs) on which the canvas is supported. The local general store has the following lengths in stock: 14, 16, 18, 20, 22, and 24 feet. You wish to make the wagon as large as possible; however, the wagon train leader has warned you that wagons in excess of 11 feet in height have blown over in some parts of the country. Which of the above lengths would you choose to make the wagon as tall as possible without exceeding 11 feet? Note that each end of the strip fits into a 6-inch rigid vertical slot.

Hints: Thin strips of wood under such large deflections obey Hook's law of linear elasticity and are closely modeled as *elastica* or *elastic lines.* These are known to have a curvature κ satisfying the nonlinear differential equation

$$\frac{d^2\kappa}{ds^2} = -\frac{1}{2}\kappa^3,$$

where s is the curve length measured along the elastica. Remember from calculus that

$$\kappa = \frac{d\theta}{ds},$$

where θ is the angle made by the curve with respect to some reference line. If we choose this reference line to be the x axis of a cartesian coordinate system, then

$$\frac{dx}{ds} = \cos\theta$$

and

$$\frac{dy}{ds} = \sin\theta.$$

Using symmetry about the vertical axis, the problem becomes a boundary-value problem which can be solved using a shooting method.

It is convenient to choose the coordinate system as shown in the figure. The fact that the ends of the strip are fixed in rigid slots gives the boundary condition $\kappa(0) = 0$. Symmetry requires that the top (center) of the strip must be horizontal, so $\theta(S) + \theta(0) = \pi/2$. There are two parameters to shoot on: $\theta(0)$ and $(d\kappa/ds)(0)$. Hence, two copies of ZEROIN are required, and one of them will have to be renamed.

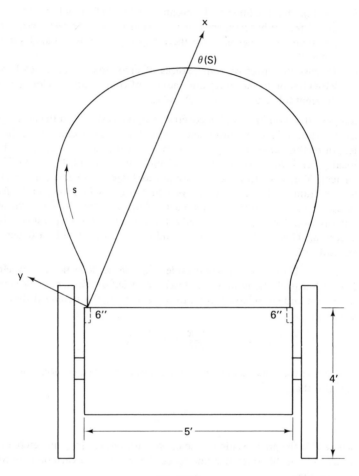

Fig. P7.8.

You can avoid repeated calls to the SIN and COS library functions by noting that

$$\frac{d}{ds}(\sin \theta) = \cos \theta \frac{d\theta}{ds} = \kappa \cos \theta,$$

$$\frac{d}{ds}(\cos \theta) = -\sin \theta \frac{d\theta}{ds} = -\kappa \sin \theta.$$

Thus, two additional differential equations can be integrated along with the other equations to compute the sines and cosines.

The object of this question is to make you wonder how people moved west without the aid of computers.

P7-9. Solve the following nonlinear boundary-value problem for $y(x)$ on the interval $0 \le x \le 1$:

$$y'' = y^2 - 1, \qquad y(0) = 0, \qquad y(1) = 1.$$

(a) Apply the shooting method described in Section 6.6, using ZEROIN and RKF45.

(b) After studying Section 7.5, try to solve the problem another way: Partition the interval into n equal subintervals:

$$0 = x_0 < x_1 < x_2 < \ldots < x_{n-1} < x_n = 1.$$

Replace the differential equation by a difference equation in $n - 1$ unknowns $y_1, y_2, \ldots, y_{n-1}$, where y_i approximates $y(x_i)$:

$$y_{i+1} - 2y_i + y_{i-1} = h^2(y_i^2 - 1), \qquad i = 1, 2, \ldots, n - 1,$$

where $y_0 = 0$ and $y_n = 1$.

Solve the *nonlinear* tridiagonal system for $n = 50$, say.

(c) For a third approach, try the following. Observe that $y'' = y^2 - 1$ can be written

$$\frac{dy}{dx}\left(\frac{(y')^2}{2} - \frac{y^3}{3} + y\right) = 0.$$

Thus,

$$\frac{(y')^2}{2} - \frac{y^3}{3} + y = c$$

for some constant c. Since $y(0) = 0$, we have $y'(0) = \sqrt{2c}$. So if we can compute c, the boundary-value problem has been converted into an initial value problem. Integrating the above equation gives

$$x = \int_0^y \frac{dy}{\sqrt{2}\left(c + \frac{y^3}{3} - y\right)^{1/2}}.$$

Now, use ZEROIN and QUANC8 to find c by solving

$$1 = \int_0^1 \frac{dy}{\sqrt{2}\left(c + \frac{y^3}{3} - y\right)^{1/2}},$$

and then use RKF45 to obtain the desired solution.

8 OPTIMIZATION

A common problem in scientific computation is to determine the maximum or minimum (and the corresponding arguments) of a real-valued function $f(x_1, \ldots, x_n)$ of n real variables over a set S in an n-dimensional space. The word *optimization* denotes either the minimization or maximization of a function. Sometimes the set S is the entire n-dimensional space; if so, the optimization problem is said to be *unconstrained*. Otherwise, the problem is *constrained* by whatever conditions define the set S. Typically S is defined by a set of nonlinear functions satisfying equality or inequality conditions. In general, a point x is in S if and only if x satisfies the inequalities

$$g_i(\mathbf{x}) \geq 0, \qquad i = 1, \ldots, m,$$

where the g_i are prescribed functions of $\mathbf{x}$.

Constrained problems where the g_i are nonlinear functions are usually considerably more difficult to solve than corresponding unconstrained problems. If the function f and all of the constraints g_i are linear functions, the problem is one of *linear programming*; it can be shown that the solution lies at a vertex of the convex polyhedron defined by the constraints in n dimensions. The usual method of solution is to search over the vertices by moving from one vertex to an adjacent vertex. The difficult problems of linear programming are principally associated with solving problems with very large n which lead to sparse matrix problems. Such problems are rendered difficult by the combinatorial complexity of a general polyhedron in n dimensions. If either the function f or any of the constraints are nonlinear, the problem is one of *nonlinear programming*. Problems of linear and nonlinear programming are beyond the scope of this book. The interested reader should consult,

178

for example, Orchard-Hays (1968) or Dantzig (1963). We shall confine our discussion to problems that are either unconstrained or constrained to an interval.

Optimization problems often arise directly in the context of finding the best design of something. For example, one may seek the values of some n parameters of a system so as to minimize the cost, which is expressed as a function of the parameters.

Sometimes optimization problems arise indirectly as a means of solving some other problem. A standard example is the reduction of the problem of solving a set of n simultaneous nonlinear equations

$$f_1(x_1, \ldots, x_n) = 0, \quad f_2(x_1, \ldots, x_n) = 0, \quad \ldots, \quad f_n(x_1, \ldots, x_n) = 0$$

to the minimization of

$$g(x_1, \ldots, x_n) = \sum_{i=1}^{n} |f_i(x_1, \ldots, x_n)|^2.$$

Obviously the zero-valued minima of g are exactly the solutions of the system. (There may also be some extraneous nonzero local minima.) However, one would not usually solve this problem by a general minimization algorithm since special advantage can be taken of the type of function being minimized.

As in the calculus of smooth functions of one variable, it is easier to find relative or local minima of a function than it is to find the absolute or global minimum over the entire domain. Indeed, there is no known practical algorithm for finding global minima for $n > 2$ or 3. Even to find an approximate global minimum over the unit cube in n dimensions would require evaluating the function over a closely packed lattice of points inside the cube and knowing a priori that the function is smooth enough not to have a *spike* in between the lattice points. But a dense lattice of points in the n-dimensional cube is far too numerous to deal with. If $n = 10$, for example, to distribute 10 abscissas in each dimension would require 10^{10} points in all. So, in practice, the only way to find a global minimum is to have information from the problem itself about the location of such a minimum and then search for the local minimum. Hence we shall confine our attention to finding local minima or maxima. Techniques for finding minima are immediately translatable into techniques for maximizing (because the minima of f are the maxima of $-f$), and we shall refer to the two problems interchangeably.

8.1. ONE-DIMENSIONAL OPTIMIZATION

Suppose that f is a real-valued function defined on [0, 1]. Suppose, moreover, that there is a unique value $\bar{x}$ such that $f(\bar{x})$ is the maximum of $f(x)$ on [0, 1] and that $f(x)$ strictly increases for $x \leq \bar{x}$ and that $f(x)$ strictly

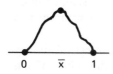 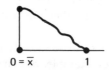 

Fig. 8.1.

decreases for $\bar{x} \leq x$. Such a function is called *unimodal*, and its graph takes one of the three forms shown in Fig. 8.1. Notice that a unimodal function need not be smooth or even continuous. It immediately follows that for any two points in the interval x_1, x_2 such that $x_1 < x_2 \leq \bar{x}$ we have $f(x_1) < f(x_2)$. Similarly, if $\bar{x} \leq x_1 < x_2$, then $f(x_1) > f(x_2)$. Conversely, if $x_1 < x_2$ and $f(x_1) < f(x_2)$, then $x_1 \leq \bar{x} \leq 1$, and if $f(x_1) > f(x_2)$, then $0 \leq \bar{x} \leq x_2$. [Of course, if $f(x_1) = f(x_2)$, we have the extra information that $x_1 \leq \bar{x} \leq x_2$, but we never use this.] The problem is to find a set of abscissas $x_1, x_2, \ldots, x_k$ at which the function f is evaluated to determine that the optimum value of f lies in the interval $x_{i-1} \leq \bar{x} \leq x_{i+1}$ for some i. Such an interval is called the *interval of uncertainty*.

The algorithm for choosing the x_i ($i = 1, \ldots, k$) is called a *search plan*. If we know only that f is unimodal, what is the optimum strategy for finding $\bar{x}$? An *optimal* search plan for a given number of function evaluations is one that results in the smallest interval of uncertainty. In some situations, such as physical experiments or some parallel processing computers, we are forced to evaluate the function f at each of the values of x_i simultaneously. It can be shown that the optimal search plan is given by

$$x_i = \frac{(1 + \epsilon)\lfloor (i + 1)/2 \rfloor}{(k/2) + 1} - \left(\left\lfloor \frac{i+1}{2} \right\rfloor - \left\lfloor \frac{i}{2} \right\rfloor \right)\epsilon,$$

where $\lfloor y \rfloor$ denotes the largest integer not greater than y. With k evaluations of f, the resulting interval of uncertainty has a length given by

$$\frac{1 + \epsilon}{(k/2) + 1}.$$

In the above formulas, ϵ is chosen just large enough so that $f(x) \neq f(x + \epsilon)$ for any point $x \in [0, 1]$ such that x is not within ϵ of $\bar{x}$. For more discussion on simultaneous searches, see Wilde (1964). In the remainder of this chapter, we shall assume that the function f will be evaluated sequentially.

Suppose that we are allowed to sequentially evaluate the function k times, where $k > 1$ is given. How can we use these evaluations to locate $\bar{x}$ with the smallest possible interval of uncertainty?

A theory was started by J. Kiefer in the early 1950s. The algorithm for an optimum strategy will produce k test points sequentially, which we choose to number x_k, x_{k-1}, ..., x_2, x_1. At the first stage two points x_k and x_{k-1} are chosen simultaneously, with $x_{k-1} < x_k$. If $f(x_{k-1}) \leq f(x_k)$, then the interval of uncertainty is reduced to $[x_{k-1}, 1]$, *and we already have the value of $f(x_k)$ as a start on our further refinement of the interval of uncertainty.* On the other hand, if $f(x_{k-1}) \geq f(x_k)$, then the interval of uncertainty is reduced to $[0, x_k]$, and we know $f(x_{k-1})$ already. The key to the optimum choice of x_{k-1} and x_k is to be sure that the one test point inherited by the reduced interval of uncertainty is itself located at an optimal position for the continued search.

It is useful to define the *coordinate r of a point x relative to an interval* $[a, b]$ as the number $r = (x - a)/(b - a)$. Thus, a has the coordinate 0 and b the coordinate 1, relative to the interval $[a, b]$, while the midpoint has the coordinate $\frac{1}{2}$.

Let $F_0 = F_1 = 1$, $F_2 = 2$, $F_3 = 3$, $F_4 = 5$, $F_5 = 8$, ..., $F_k = F_{k-1} + F_{k-2}$. The F_i are the famous Fibonacci numbers. The optimum strategy for the sequential search for a maximum is called a *Fibonacci search* because it is intimately related to these numbers. In the optimum strategy one picks $x_{k-1} = F_{k-2}/F_k$ and $x_k = F_{k-1}/F_k$. Whichever interval $[0, x_k]$ or $[x_{k-1}, 1]$ becomes the reduced interval of uncertainty, the inherited point will be located at one of the following two coordinates relative to the new interval: F_{k-3}/F_{k-1} or F_{k-2}/F_{k-1}. Then x_{k-2} is selected to be the other one of the above two coordinates relative to the new interval. From $f(x_{k-2})$ and the function value inherited from the first interval, one can further reduce the interval of uncertainty and inherit one function value.

At the last stage, one has some interval of uncertainty $[a, b]$ with its midpoint $\frac{1}{2}$ as the inherited point. Then x_1 is selected to have relative coordinate $\frac{1}{2} + \epsilon$, and the final interval of uncertainty is either $[0, \frac{1}{2} + \epsilon]$ or $[\frac{1}{2}, 1]$ relative to $[a, b]$.

At the first stage the length of the interval of uncertainty is reduced from 1 to F_{k-1}/F_k. At succeeding stages the intervals are reduced by the factors

$$\frac{F_{k-2}}{F_{k-1}}, \frac{F_{k-3}}{F_{k-2}}, \ldots, \frac{F_2}{F_3}, \frac{F_1}{F_2}(1 + 2\epsilon).$$

Thus the length of the final interval of uncertainty is $(1 + 2\epsilon)/F_k$. Ignoring ϵ, we note that $1/F_k$ is asymptotically equal to r^k, as $k \longrightarrow \infty$, where

$$r = \frac{\sqrt{5} - 1}{2} \approx 0.6180.$$

Thus, asymptotically for large k, each step of the Fibonacci search reduces the interval of uncertainty by the factor 0.6180. This should be compared with 0.5, the reduction of the interval of uncertainty of the bisection method for finding a zero of a function.

For large values of k, the locations of x_{k-1} and x_k are close to $1 - r \approx$ 0.3820 and $r \approx 0.6180$, respectively, and starting with these values is close to an optimum strategy. To see how to proceed further, suppose (for example) that $f(0.3820) > f(0.6180)$. Then $\bar{x}$ is known to be in the interval $[0, 0.6180]$. Hence one wants to evaluate f at the points $0.3820*0.6180$ and $0.6180*0.6180$. But, since $0.6180*0.6180 \approx 0.3820 \approx x_{k-1}, f$ is already known there. Thus, again, only one evaluation of f is required at each stage of the iteration after the first, and each stage reduces the length of the interval of uncertainty by the factor 0.6180. In contrast with the Fibonacci search, one does not have to fix the number k before the search begins.

Because of the history of the number $r \approx 0.6180$, this method of finding $\bar{x}$ is called the *golden-section search*. The number $1/r = \phi = (1 + \sqrt{5})/2 = 1.6180 \ldots$ is called the *golden ratio*. In Chapter 7, we noted that the rate of convergence for the secant method is equal to ϕ. The golden ratio ϕ is a fundamental ratio that pops up in many different places. For an interesting discussion of ϕ, see Gardner (1961).

The golden-section search is analogous to the method of bisection for finding a real zero of a function (see Chapter 7) in that it is guaranteed to work in the worst possible case, and the price of this safety is slowness of convergence, which is only linear. Golden-section search takes no advantage whatever of the possible smoothness of f.

If f is known to have several continuous derivatives and if one can start close enough to $\bar{x}$, then an iterative process can be set up to find $\bar{x}$ as follows: Start with three arbitrary real numbers v_1, v_2, v_3. At the general stage one has v_{k-2}, v_{k-1}, and v_k. Let v_{k+1} be the abscissa of the maximum ordinate of the parabola (with vertical axis) through $(v_i, f(v_i))$, $i = k - 2, k - 1, k$. Continue the iteration with v_{k-1}, v_k, and v_{k+1}. This iteration is called *successive parabolic interpolation*. For good enough starts, the iteration can be proved to converge with convergence of order $\approx 1.324 \ldots$, provided $f''(x) > 0$ at the minimum [that is, $f'(x)$ has a simple zero].

8.2. SUBROUTINE FMIN

Brent (1973) has created a minimum finder which uses a combination of the golden-section search and successive parabolic interpolation. The algorithm is entirely analogous to the zero finder ZEROIN of Chapter 7, which uses a combination of bisection and inverse quadratic interpolation. A Fortran translation of Brent's Algol 60 algorithm appears at the end of this section. It is a function subroutine

FUNCTION FMIN (A, B, F, TOL).

Here $[a, b]$ is the initial interval on which the function f is defined. TOL is an input parameter that, roughly speaking, gives the desired length of the interval of uncertainty on exit. Thus, if you set TOL to 10^{-10}, and the result returned by FMIN has a magnitude of about 1, then the answer has about ten significant digits. If you set TOL to 10^{-6} and the answer is (say) 0.00000385746, then the result probably has no significant digits, but if you interpret a result of this magnitude as zero, then you do not need any significant digits. FMIN uses the golden-section search, switching when possible to successive parabolic interpolation.

The subroutines ZEROIN and FMIN have the same parameters—an interval to be searched, a function to be evaluated, and a tolerance. Both subroutines attempt to reduce the length of the interval until it is less than the tolerance. However, there is an important difference between the two routines which affects the choice of the tolerance: If $f(x) = 0$ and $f'(x) \neq 0$, then for small ϵ

$$f(x + \epsilon) = f(x) + \epsilon f'(x) + \epsilon^2 \frac{f''(x)}{2} + \cdots$$

$$\approx c\epsilon,$$

where $c = f'(x)$.

Small changes in x cause proportionally small changes in $f(x)$. It is reasonable to choose as the tolerance for ZEROIN a value roughly the size of the errors in the function values. Often this can be as small as the machine roundoff level.

However, if we are seeking a minimum where $f'(x) = 0$ and $f''(x) \neq 0$, then for small ϵ,

$$f(x + \epsilon) = f(x) + \epsilon^2 \frac{f''(x)}{2} + \cdots$$

$$\approx f(x) + c\epsilon^2,$$

where $c = f''(x)/2$.

A change of order ϵ in x now causes a change of order ϵ^2 in $f(x)$. So it is unreasonable to choose as the tolerance for FMIN a value smaller than the square root of the error in the function values. In other words, simple zeros of a function can often be found to nearly *full* machine precision, but points where a function obtains a minimum may be found to only about *half* precision.

If TOL is the tolerance input to ZEROIN or FMIN, and MACHEPS is the machine precision (computed by the subroutines), then ZEROIN never evaluates the function at points closer together than

$$2*\text{MACHEPS}*\text{ABS(X)} + \text{TOL}/2,$$

while FMIN never evaluates the function at points closer together than

$$\text{SQRT(MACHEPS)}*\text{ABS(X)} + \text{TOL}/3.$$

Thus small values of TOL are much more likely to be overridden by FMIN than by ZEROIN.

Unimodal functions occur relatively infrequently in practice. One often knows only that $f(x)$ is decreasing, as x increases from 0 to some unknown minimum point $\bar{x}$, after which $f(x)$ increases. One wishes to locate an approximation to $\bar{x}$ as rapidly as possible, without any prior idea of where it might be located. We know of no theory for this problem. Programs sometimes pick some $a > 0$ and then evaluate $f(0)$, $f(a)$, $f(2a)$, $f(2^2a)$, $f(2^3a)$, etc., as long as the values continue to decrease, accepting as an approximate $\bar{x}$ the 2^ka with the least $f(x)$. Or, if $f(a) > f(0)$, the program evaluates $f(2^{-1}a)$, $f(2^{-2}a), \dots$ until a value $f(2^{-r}a)$ is found such that $f(2^{-r}a) < f(0)$, and then $2^{-r}a$ is accepted as an approximate $\bar{x}$. The reader can devise various ways of refining the $\bar{x}$ accepted by the above crude methods. In many applications, however, it is well to accept the crude value of $\bar{x}$ and then return to the main algorithm of which the one-dimensional search is a subroutine.

For a sample program using FMIN, we use the same function as we used for ZEROIN, $f(x) = x^3 - 2x - 5$. Since $f'(x)$ is a quadratic, the extrema can be found analytically. Note that the value of TOL used here is the square root of the value used with ZEROIN.

The output is $Z = 0.81650$.

```
C   SAMPLE PROGRAM FOR FMIN
C
    REAL FUNCTION F(X)
    REAL X
    F = X*(X*X - 2.) - 5.
    RETURN
    END
C
    EXTERNAL F
    REAL A,B,Z,TOL,FMIN
    A = 0.
    B = 1.
    TOL = 1.0E-5
    Z = FMIN(A,B,F,TOL)
    WRITE(6,1) Z
1   FORMAT(3H Z=,F12.5)
    STOP
    END
```

```
      REAL FUNCTION FMIN(AX,BX,F,TOL)
      REAL AX,BX,F,TOL
C  µ
C  ˜    AN APPROXIMATION  X  TO THE POINT WHERE  F  ATTAINS A MINIMUM  ON
C THE INTERVAL  (AX,BX)  IS DETERMINED.
C
C
C INPUT..
C
C AX    LEFT ENDPOINT OF INITIAL INTERVAL
C BX    RIGHT ENDPOINT OF INITIAL INTERVAL
C F     FUNCTION SUBPROGRAM WHICH EVALUATES  F(X)  FOR ANY  X
C       IN THE INTERVAL  (AX,BX)
C TOL   DESIRED LENGTH OF THE INTERVAL OF UNCERTAINTY OF THE FINAL
C       RESULT ( .GE. 0.0)
C
C
C OUTPUT..
C
C FMIN  ABCISSA APPROXIMATING THE POINT WHERE  F  ATTAINS A MINIMUM
C
C
C      THE METHOD USED IS A COMBINATION OF  GOLDEN  SECTION  SEARCH  AND
C SUCCESSIVE PARABOLIC INTERPOLATION.  CONVERGENCE IS NEVER MUCH SLOWER
C THAN  THAT  FOR  A  FIBONACCI SEARCH.  IF  F  HAS A CONTINUOUS SECOND
C DERIVATIVE WHICH IS POSITIVE AT THE MINIMUM (WHICH IS NOT  AT  AX  OR
C BX),  THEN  CONVERGENCE  IS  SUPERLINEAR, AND USUALLY OF THE ORDER OF
C ABOUT  1.324....
C      THE FUNCTION  F  IS NEVER EVALUATED AT TWO POINTS CLOSER TOGETHER
C THAN  EPS*ABS(FMIN) + (TOL/3), WHERE EPS IS  APPROXIMATELY THE SQUARE
C ROOT OF  THE  RELATIVE  MACHINE  PRECISION.   IF   F   IS A UNIMODAL
C FUNCTION AND THE COMPUTED VALUES OF  F    ARE  ALWAYS  UNIMODAL  WHEN
C SEPARATED BY AT LEAST  EPS*ABS(X) + (TOL/3), THEN   FMIN  APPROXIMATES
C THE ABCISSA OF THE GLOBAL MINIMUM OF  F  ON THE INTERVAL AX,BX  WITH
C AN ERROR LESS THAN  3*EPS*ABS(FMIN) + TOL.  IF  F   IS NOT UNIMODAL,
C THEN FMIN MAY APPROXIMATE A LOCAL, BUT PERHAPS NON-GLOBAL, MINIMUM TO
C THE SAME ACCURACY.
C      THIS FUNCTION SUBPROGRAM IS A SLIGHTLY MODIFIED  VERSION  OF  THE
C ALGOL  60 PROCEDURE  LOCALMIN  GIVEN IN RICHARD BRENT, ALGORITHMS FOR
C MINIMIZATION WITHOUT DERIVATIVES, PRENTICE - HALL, INC. (1973).
C
C
      REAL  A,B,C,D,E,EPS,XM,P,Q,R,TOL1,TOL2,U,V,W
      REAL  FU,FV,FW,FX,X
C
C C IS THE SQUARED INVERSE OF THE GOLDEN RATIO
C
      C = 0.5*(3. - SQRT(5.0))
```

```
C
C  EPS IS APPROXIMATELY THE SQUARE ROOT OF THE RELATIVE MACHINE
C  PRECISION.
C
      EPS = 1.00
   10 EPS = EPS/2.00
      TOL1 = 1.0 + EPS
      IF (TOL1 .GT. 1.00) GO TO 10
      EPS = SQRT(EPS)
C
C  INITIALIZATION
C
      A = AX
      B = BX
      V = A + C*(B - A)
      W = V
      X = V
      E = 0.0
      FX = F(X)
      FV = FX
      FW = FX
C
C  MAIN LOOP STARTS HERE
C
   20 XM = 0.5*(A + B)
      TOL1 = EPS*ABS(X) + TOL/3.0
      TOL2 = 2.0*TOL1
C
C  CHECK STOPPING CRITERION
C
      IF (ABS(X - XM) .LE. (TOL2 - 0.5*(B - A))) GO TO 90
C
C  IS GOLDEN-SECTION NECESSARY
C
      IF (ABS(E) .LE. TOL1) GO TO 40
C
C  FIT PARABOLA
C
      R = (X - W)*(FX - FV)
      Q = (X - V)*(FX - FW)
      P = (X - V)*Q - (X - W)*R
      Q = 2.00*(Q - R)
      IF (Q .GT. 0.0) P = -P
      Q =  ABS(Q)
      R = E
      E = D
C
C  IS PARABOLA ACCEPTABLE
C
   30 IF (ABS(P) .GE. ABS(0.5*Q*R)) GO TO 40
      IF (P .LE. Q*(A - X)) GO TO 40
      IF (P .GE. Q*(B - X)) GO TO 40
C
C  A PARABOLIC INTERPOLATION STEP
C
      D = P/Q
      U = X + D
```

```
C
C  F MUST NOT BE EVALUATED TOO CLOSE TO AX OR BX
C
      IF ((U - A) .LT. TOL2) D = SIGN(TOL1, XM - X)
      IF ((B - U) .LT. TOL2) D = SIGN(TOL1, XM - X)
      GO TO 50
C
C  A GOLDEN-SECTION STEP
C
   40 IF (X .GE. XM) E = A - X
      IF (X .LT. XM) E = B - X
      D = C*E
C
C  F MUST NOT BE EVALUATED TOO CLOSE TO X
C
   50 IF (ABS(D) .GE. TOL1) U = X + D
      IF (ABS(D) .LT. TOL1) U = X + SIGN(TOL1, D)
      FU = F(U)
C
C  UPDATE  A, B, V, W, AND X
C
      IF (FU .GT. FX) GO TO 60
      IF (U .GE. X) A = X
      IF (U .LT. X) B = X
      V = W
      FV = FW
      W = X
      FW = FX
      X = U
      FX = FU
      GO TO 20
   60 IF (U .LT. X) A = U
      IF (U .GE. X) B = U
      IF (FU .LE. FW) GO TO 70
      IF (W .EQ. X) GO TO 70
      IF (FU .LE. FV) GO TO 80
      IF (V .EQ. X) GO TO 80
      IF (V .EQ. W) GO TO 80
      GO TO 20
   70 V = W
      FV = FW
      W = U
      FW = FU
      GO TO 20
   80 V = U
      FV = FU
      GO TO 20
C
C  END OF MAIN LOOP
C
   90 FMIN = X
      RETURN
      END
```

8.3. OPTIMIZATION IN MANY DIMENSIONS

The local minimization of a function of n variables is so important that algorithms have been devised for it for over 130 years. An old method, known now as the *method of steepest descent*, was proposed by A. Cauchy in 1845. Let the vector $(x_1, \ldots, x_n)^T$ be denoted by $\mathbf{x}$, and let the function $f(\mathbf{x})$ be assumed to have continuous partial derivatives of several orders. Let $\mathbf{g}(\mathbf{x})$ denote the gradient of f at $\mathbf{x}$—i.e., the vector whose ith component is $\partial f/\partial x_i$. Then $-\mathbf{g}(\mathbf{x})$ defines the direction of steepest descent of f at $\mathbf{x}$. Cauchy proposes searching the half line defined by $\mathbf{x} - t\mathbf{g}$ $(0 < t < \infty)$ for a minimum value of f. This is a one-dimensional search of the type we discussed at the end of Section 8.1. Having found this minimum, one starts over and searches along the half line of steepest descent from the new $\mathbf{x}$. Under weak hypotheses, this method will converge to the local minimum of f in whose *basin* the first $\mathbf{x}$ is located.

Theoretical analysis has predicted that Cauchy's method will converge extremely slowly in some cases, and experience has confirmed this in many practical cases, even for such modest values of n as 2, 3, or 4. The reason is easily seen for $n = 2$ if the function f is constant on ellipses of large eccentricity, as in Fig. 8.2. (Such a function might be $x_1^2 + 100,000x_2^2$.) The path to the minimum at the center proceeds very slowly by a route that gradually works down the valley, oscillating back and forth in the local gradient directions. There is great need to accelerate the convergence.

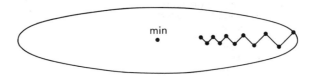

Fig. 8.2.

The big difficulty is that the local gradient does not even approximately point toward the center of the ellipses. To find the direction for rapid progress for a general strictly quadratic function f in n variables, let us expand $f(\mathbf{x})$ about the minimum point, which we denote by $\mathbf{a}$. Let B be the constant matrix of second partial derivatives of f at $\mathbf{a}$—the so-called *Hessian matrix*:

$$b_{i,j} = \frac{\partial^2 f}{\partial x_i \, \partial x_j}.$$

Then, by Taylor's theorem, since $\mathbf{g}(\mathbf{a}) = 0$ at a minimum, we can write

$$f(\mathbf{x}) = f(\mathbf{a}) + \tfrac{1}{2}(\mathbf{x} - \mathbf{a})^T B(\mathbf{x} - \mathbf{a}).$$

Thus, the gradient **g** at **x** is given by

$$\mathbf{g(x)} = B(\mathbf{x} - \mathbf{a}).$$

If we are at an arbitrary point **x**, we can compute **a** by

$$\mathbf{a} = \mathbf{x} - B^{-1}\mathbf{g(x)}.$$

Thus the minimum of f can be found from a knowledge of B^{-1} and the gradient direction **g** at **x**. (B is a positive-definite matrix.) Let the matrix B^{-1} be denoted by H. Then the matrix H is the operator which turns the local direction of steepest descent $-\mathbf{g}$ at **x** into the true direction from **x** to the center of the ellipsoids, i.e., to the minimum point of f. In terms of H, the minimum **a** is given by

$$\mathbf{a} = \mathbf{x} - H\mathbf{g}.$$

One can look at this as representing the Newton method for solving the system of equations $\mathbf{g(x)} = 0$.

If a smooth function f is not quadratic, it will nevertheless be approximately quadratic in the neighborhood of a minimum at which the Hessian matrix does not vanish. Hence the previous equations suggest an attack on the problem of finding the minimum of any f. We shall use an iteration of the form

$$\mathbf{x}_{k+1} = \mathbf{x}_k - \alpha_k H_k \mathbf{g}_k,$$

where $\mathbf{g}_k = \mathbf{g(x}_k)$ and H_k is a kth approximation to the inverse H of the Hessian matrix B of the function f at **a**, the minimum. The positive scalar α_k is determined at each step of the iteration such that a local minimum of f is achieved along the direction $-H_k \mathbf{g}_k$ from $\mathbf{x}_k$. The iteration tries to discover the minimum point **a** and the inverse Hessian matrix H simultaneously.

The basic idea is to build up H_k in such a way that at the nth stage H_n would equal H if f were a quadratic function. Thus, if f were quadratic, the nth iterate $\mathbf{x}_n$ would equal **a**, using exact arithmetic. This suggests that for nonquadratic functions, the nth iterate $\mathbf{x}_n$ should be a much closer approximation to the minimum **a** than $\mathbf{x}_0$.

Methods of this type are called *variable metric* methods, because the matrix B can be interpreted as a metric in the space of **g**. Another common name is *quasi-Newton*. An early algorithm was devised by Davidon in 1959. There have been many improvements of quasi-Newton methods since Davidon's original paper. The reader is referred to a survey by Broyden found in Murray (1972).

M. J. D. Powell has proved the convergence of the algorithm for any $f(\mathbf{x})$ for which the Hessian matrix is a positive-definite matrix for all **x**;

however, one rarely uses the algorithm for functions satisfying this condition. The rate of convergence is known to be faster than linear.

Algorithms of this type are called *quadratically terminating* because they converge in a finite number of steps for quadratic functions. This does not imply that the degree of convergence is 2 for general functions, although it may be.

Subroutines for nonlinear optimization can be classified as to whether they handle unconstrained or constrained problems, whether the objective function is a sum of squares or a general nonlinear function, and whether or not evaluation of derivatives is required. Collections of these routines also usually include programs for the solution of simultaneous nonlinear equations.

Much of the recent development of reliable, efficient, general-purpose subroutines has taken place in Great Britain, particularly at the Atomic Energy Research Establishment, Harwell, and the National Physical Laboratory, Teddington. Examples of such routines are the Algol 60 procedures of Gill et al. (1972). Many other routines are described along with the pertinent theoretical background in two companion surveys, one edited by Murray (1972) on unconstrained optimization and the other edited by Gill and Murray (1974) on constrained optimization.

At the time this is being written in 1976, the most up-to-date routines are those available from Harwell, Teddington, and the British Numerical Algorithms Group, Oxford. Outside Great Britain, a group at the U.S. Argonne National Laboratory is developing a collection of subroutines known as MINPACK. It will be made publically available after careful testing.

PROBLEMS

P8-1. When the definition of erf (x) is extended to complex arguments, one of the terms involved is a real function of a real variable known as Dawson's integral,

$$D(x) = e^{-x^2} \int_0^x e^{t^2} \, dt.$$

Since $D(x)$ approaches 0 as x approaches either 0 or ∞ and $D(x)$ is positive in between, it must have a finite maximum. Use FMIN to find the maximum of $D(x)$.

P8-2. Use FMIN to find the maximum of the function

$$f(x) = (\sin x)^6 \tan (1 - x)e^{30x}$$

on the interval [0, 1].

P8-3. For any $\lambda > 0$, let $z(\lambda)$ be the least positive zero of the function $y(t)$ that solves the ordinary differential equation problem

$$y''(t) + I_0(y) + \frac{t}{10} = 0,$$

$$y(0) = 0, \qquad y'(0) = \lambda,$$

where $I_0(t)$ is the zeroth-order modified Bessel function [see Abramowitz and Stegun (1964), pp. 374–375].

The graph of the solution $y(t)$ is approximately as shown here.

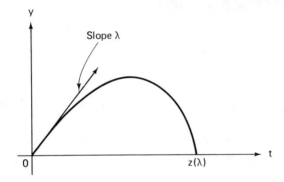

Fig. P8.3.

(a) Find λ_{max}, the unique value of λ such that $z(\lambda)$ is a maximum.
(b) Also find $z(\lambda_{max})$.
(c) Give a table of values of $y(t)$ for $t = 0(0.1)z(\lambda_{max})$ for the $y(t)$ which maximizes $z(\lambda)$, i.e., for $y(t)$ satisfying $y'(0) = \lambda_{max}$.
(d) Give a discussion of the accuracy of your results.

9
LEAST SQUARES AND THE SINGULAR VALUE DECOMPOSITION

9.1. LEAST-SQUARES DATA FITTING

The following problem occurs in many different branches of science. Suppose we are given m data points

$$(t_i, y_i), \qquad i = 1, \ldots, m.$$

We think of t as the independent variable and y as the dependent variable satisfying some unknown functional relationship

$$y_i = y(t_i).$$

Suppose further that y is to be approximated by a linear combination of n given basis functions ϕ_j,

$$y(t) \approx c_1 \phi_1(t) + c_2 \phi_2(t) + \ldots + c_n \phi_n(t).$$

The linear combination on the right is referred to as a *linear mathematical model*. The problem is to choose the n coefficients $c_1, \ldots, c_n$ so that the model fits the data in some sense. The model is linear because the coefficients appear linearly, although the basis functions may be nonlinear functions of t.

The most common linear model is a polynomial

$$y(t) \approx c_1 + c_2 t + c_3 t^2 + \ldots + c_n t^{n-1}.$$

This is achieved by taking $\phi_j(t) = t^{j-1}$, although, as we shall see, other basis polynomials may be more appropriate. Other examples include trigonometric approximation,

$$y(t) \approx c_1 \sin t + c_2 \sin 2t + \ldots + c_n \sin nt,$$

where $\phi_j(t) = \sin jt$, and exponential approximation with fixed exponents,

$$y(t) \approx c_1 e^{\lambda_1 t} + \ldots + c_n e^{\lambda_n t},$$

where $\phi_j(t) = \exp(\lambda_j t)$ for given λ_j. (If the λ_j are also to be determined, this becomes a nonlinear model. See Section 8.5 and Problem P9-3.)

We shall consider the case where m, the number of data points, is greater than or equal to n, the number of unknown coefficients. Then the problem of choosing the coefficients is overdetermined, and it is usually not possible to have the model exactly fit, that is, interpolate, the data.

Of the many different criteria for determining the coefficients c_j, the method of *least squares* is used most frequently. For any choice of the c_j, the residual at the ith data point is

$$r_i = \sum_{j=1}^{n} c_j \phi_j(t_i) - y_i.$$

The least-squares criterion specifies that the c_j be chosen to minimize the sum of the squares of the residuals, that is,

$$\underset{c_j}{\text{minimize}} \sum_{i=1}^{m} r_i^2.$$

If the model can exactly fit the data, the minimum will be zero, so interpolation is included as a special case.

The least-squares criterion does not necessarily determine a unique set of coefficients. If the basis functions are linearly dependent at the data points, which means that there are nonzero coefficients γ_j for which

$$\sum_{j=1}^{n} \gamma_j \phi_j(t_i) = 0, \qquad i = 1, \ldots, m,$$

then any multiple of the γ_j's can be added to the c_j's without changing the sum of the squares of the residuals. An important and often overlooked task in least-squares data fitting with general basis functions is the detection and proper handling of such dependence and nonuniqueness.

There are many different algorithms for computing a set of coefficients which give a minimum sum of squares. One possibility is to use calculus. Let

$$r = \left(\sum_{i=1}^{m} \left(\sum_{j=1}^{n} c_j \phi_j(t_i) - y_i \right)^2 \right)^{1/2}.$$

Our goal is to minimize r, or, equivalently, to minimize r^2, so we require for $k = 1, \ldots, n$

$$\frac{\partial r^2}{\partial c_k} = 0.$$

Taking the derivatives and interchanging orders of summation, we obtain

$$\sum_{j=1}^{n} \left(\sum_{i=1}^{m} \phi_j(t_i)\phi_k(t_i) \right) c_j = \sum_{i=1}^{m} y_i \phi_k(t_i).$$

This is a set of n simultaneous linear equations in the n unknown c_j's. It can be written in matrix form

$$Pc = q,$$

where

$$p_{kj} = \sum_{i=1}^{m} \phi_k(t_i)\phi_j(t_i),$$

$$q_k = \sum_{i=1}^{m} \phi_k(t_i)y_i.$$

For example, with least-squares approximation by a straight line,

$$y(t) \approx c_1 + c_2 t,$$

$$P = \begin{pmatrix} m & \sum t_i \\ \sum t_i & \sum t_i^2 \end{pmatrix},$$

$$q = \begin{pmatrix} \sum y_i \\ \sum t_i y_i \end{pmatrix}.$$

The equations $Pc = q$ are called the *normal equations*. Notice that the matrix P depends only on the basis functions; it does not involve the values of y_i. In principle, the normal equations could be solved using DECOMP and SOLVE from Chapter 3. However, since $p_{kj} = p_{jk}$, the matrix P is symmetric, and the time and storage required by DECOMP can be cut in half. Moreover, P can be shown to be *positive definite*, and so no pivoting is required. Consequently, by using a variant of Gaussian elimination intended for positive-definite, symmetric matrices, the normal equations can be solved with less than half the effort required by DECOMP and SOLVE.

However, there are serious fundamental drawbacks to the use of the normal equations. It turns out that the matrix P often has a very high condition number, so that no matter how the normal equations are actually solved, errors in the data and roundoff errors introduced during the solution are excessively magnified in the computed coefficients.

In the extreme situation where the basis functions $\phi_j(t)$ are linearly dependent, it can be shown that P is singular, and the condition number can be regarded as infinite. Consequently, methods which avoid the high condition number inherent in the normal equations are also methods which can detect linear dependence among the basis functions. Gaussian elimination and its variants are not well suited to detecting such linear dependence. Our condition estimator is some help, but it can only warn of trouble—it cannot suggest a cure.

The most reliable method for computing the coefficients for general least-squares problems is based on a matrix factorization known as the singular value decomposition. There are other methods which require less computer time and storage, but they are less effective in dealing with errors in the data, roundoff errors, and linear dependence.

The singular value decomposition, or SVD, is described in detail in later sections of this chapter. However, it is not necessary to understand these sections in detail to use the SVD in least-squares data fitting.

The SVD approach begins with a matrix which is known in the statistical analysis of experiments as the *design matrix*. It is the rectangular matrix A with m rows and n columns whose elements are

$$a_{ij} = \phi_j(t_i).$$

If y denotes the m vector with elements y_i and c denotes the n vector with components c_j, then the approximate equations

$$\sum_{j=1}^{n} c_j\phi_j(t_i) \approx y_i, \qquad i = 1, \ldots, m,$$

can be written as

$$Ac \approx y.$$

The SVD subroutine given at the end of the chapter takes A as input and returns three matrices Σ, U, and V as output. The matrix Σ is diagonal with nonnegative diagonal entries, which are known as the *singular values* of A. The matrices U and V are used to transform the equations

$$Ac \approx y$$

into an equivalent diagonal set of equations

$$\Sigma \bar{c} \approx \bar{y}$$

Let $\sigma_j, j = 1, \ldots, n$, be the diagonal entries of Σ. In principle, if none of the σ_j are zero, the transformed equations could be solved by setting

$$\bar{c}_j = \frac{\bar{y}_j}{\sigma_j}, \qquad j = 1, \ldots, n.$$

However, this is not always desirable in practice if some σ_j are small.

It turns out that the σ_j are nonzero if and only if the basis functions ϕ_j are linearly independent at the data points. So the key to the proper use of the SVD involves a tolerance τ which reflects the accuracy of the original data and the accuracy of the floating-point arithmetic being used. Any σ_j greater than τ is acceptable, and the corresponding $\bar{c}_j$ is computed as $\bar{y}_j/\sigma_j$. Any σ_j less than τ is regarded as negligible, and the corresponding $\bar{c}_j$ could be given an arbitrary value. This arbitrary value is associated with the non-uniqueness of the set of least-squares coefficients. Changes in the data and roundoff errors less than τ can result in an entirely different set of coefficients c_j which satisfy the least-squares criterion. Since it is generally desirable to also have the coefficients as small as possible, we choose to set $\bar{c}_j = 0$ whenever $\sigma_j \leq \tau$.

The ratio $\sigma_{max}/\sigma_{min}$, where σ_{max} is the largest singular value and σ_{min} is the smallest nonzero singular value, can be regarded as a condition number of the matrix A. It is not the same condition number we discussed in Chapter 3, but it has many of the same properties and is usually about the same order of magnitude numerically. As we shall see in the next section, it is a better measure of condition for least-squares problems.

Neglecting σ_j's less than τ has the effect of decreasing the condition number to σ_{max}/τ. Since the condition number is an error magnification factor, this results in a more reliable determination of the c_j. The cost of this increased reliability is a possible increase in the size of the residuals.

The following segments of a Fortran program illustrate the use of SVD for least-squares data-fitting problems. We have not given a complete subroutine because we expect these segments to be merged with other code which reads the data, generates the basis functions, uses the coefficients to evaluate the model at other points, and so on.

The design matrix might be generated as follows:

```
      DO 20 I = 1, M
         T(I) = abscissa of I-th data point
         Y(I) = ordinate of I-th data point
         DO 10 J = 1, N
            A(I,J) = J-th basis function evaluated at T(I)
   10    CONTINUE
   20 CONTINUE.
```

In the case of polynomial approximation, an efficient way of setting up A is

```
      DO 20 I = 1, M
         T(I) = ...
         Y(I) = ...
         A(I,1) = 1.0
         DO 10 J = 2, N
            A(I,J) = T(I)*A(I,J-1)
   10    CONTINUE
   20 CONTINUE.
```

This can be immediately followed by the call to SVD. (See the comments in the listing of SVD for details of its use.)

```
CALL SVD(MDIM, M, N, A, SIGMA, .TRUE., U,
         .TRUE., V, IERR, WORK)
IF (IERR .NE. 0) WRITE(6,13)
13  FORMAT('ERROR RETURN FROM SVD')
```

We have never seen the error condition result from a proper use of SVD.

The next segment of code finds the largest singular value, which will be used in checking for negligible singular values. It also sets the coefficient vector to zero, which would be the final result if all singular values were negligible.

```
SIGMA1 = 0.
DO 30 J = 1, N
   IF (SIGMA(J) .GT. SIGMA1) SIGMA1 = SIGMA(J)
   C(J) = 0.
30  CONTINUE
```

Now, a relative error tolerance, RELERR, must be supplied. For example, if the data are accurate to three significant figures, then $RELERR = 10^{-3}$ is appropriate. If the data are regarded as exact, then RELERR should reflect the roundoff error expected in the computations within SVD itself, which is something like n times the machine epsilon. The absolute error tolerance τ is then obtained by

$$TAU = RELERR*SIGMA1.$$

The next segment applies the transformation U to the original data, solves for the transformed coefficients, and then applies the transformation V to obtain the coefficients themselves.

```
DO 60 J = 1, N
   IF (SIGMA(J) .LE. TAU) GO TO 60
   S = 0.
   DO 40 I = 1, M
      S = S + U(I,J)*Y(I)
40     CONTINUE
   S = S/SIGMA(J)
   DO 50 I = 1, N
      C(I) = C(I) + S*V(I,J)
50     CONTINUE
60  CONTINUE
```

Notice that M is the number of elements in Y and the number of rows in A and U, whereas N is the number of elements in C and the number of rows of V. All the matrices have N columns.

The coefficients are now ready for use. They could be printed out if desired.

```
                    WRITE(6, . . .) (C(J), J=1, N)
```

To evaluate the model at any point TT.

```
              YY = 0.
              DO 70 J = 1, N
                  YY = YY + C(J)*(J-th basis function
                                      evaluated at TT)
          70  CONTINUE
```

For polynomial models, Horner's scheme (see Section 4.2) should be used.

```
              YY = 0.
              DO 70 JB = 1, N
                  J = N + 1 - JB
                  YY = TT*YY + C(J)
          70  CONTINUE
```

The square root of the sum of the squares of the residuals, which is the quantity being minimized, can be easily obtained.

```
              RSQ = 0.
              DO 90 I = 1, M
                  RI = 0.
                  DO 80 J = 1, N
                      RI = RI + C(J)*A(I,J)
          80      CONTINUE
                  RSQ = RSQ + (RI − Y(I))**2
          90  CONTINUE
              R = SQRT(RSQ)
```

Finally, we have claimed that an important feature of the SVD is its ability to detect dependence and lack of uniqueness. The following code checks for negligible singular values. For each one found, a set of coefficients, called *null coefficients*, is printed out. Any multiple, α, of these coefficients can be added to the coefficients printed above without increasing r by more than $\alpha\tau$. The coefficients are normalized so that the sum of their squares is 1.0.

```
                  UNIQUE = .TRUE.
                  DO 100 J = 1, N
                      IF (SIGMA(J) .GT. TAU) GO TO 100
                      UNIQUE = .FALSE.
                      WRITE(6,101)
                      WRITE(6, . . .) (V(I,J), I = 1, N)
          100 CONTINUE
                  IF (UNIQUE) WRITE(6,102)
          101 FORMAT('NULL COEFFICIENTS')
          102 FORMAT('COEFFICIENTS ARE UNIQUE')
```

This type of auxiliary information about least-squares problems is known as *singular value analysis*. It can be very valuable in analysis of complicated mathematical models. The book by Lawson and Hanson (1974) has a complete discussion of singular value analysis.

The use of singular value analysis for least-squares-fitting problems can be illustrated by the U.S. Census data from Problem P4-5. There are $m = 8$ data points. The values of t_i are 1900, 1910, . . . , 1970, and the corresponding values of y_i, in units of a million people, are about 75.99, 91.97, . . . , 203.21. When these are fit by a quadratic

$$y(t) \approx c_1 + c_2 t + c_3 t^2$$

and the computation is done in long precision on the IBM 360, the coefficients are found to be about

$$c_1 = 0.373 \times 10^5, \qquad c_2 = -0.402 \times 10^2, \qquad c_3 = 0.108 \times 10^{-1}.$$

When the same computation is done in short precision, the coefficients are about

$$c_1 = -0.372 \times 10^5, \qquad c_2 = 0.368 \times 10^2, \qquad c_3 = -0.905 \times 10^{-2}.$$

The signs of the two sets of coefficients do not even agree. In fact, when the model is used to predict the population in 1980, the coefficients obtained with long precision predict 227.78 million, whereas the coefficients obtained with short precision predict 145.21 million. The second set of coefficients are clearly useless, but what about the first set? How reliable are they?

The design matrix for this example is 8 by 3,

$$A = \begin{pmatrix} 1 & 0.190 \times 10^4 & 0.36100 \times 10^7 \\ 1 & 0.191 \times 10^4 & 0.36481 \times 10^7 \\ \vdots & \vdots & \vdots \\ 1 & 0.197 \times 10^4 & 0.39204 \times 10^7 \end{pmatrix}$$

Its singular values can be obtained to sufficient accuracy with either precision. They are

$$\sigma_1 = 0.106 \times 10^8, \qquad \sigma_2 = 0.648 \times 10^2, \qquad \sigma_3 = 0.346 \times 10^{-3}.$$

The condition number is $\sigma_1/\sigma_3 = 0.306 \times 10^{11}$, which is the signal that there is some difficulty. For t between 1900 and 1970, the three basis functions 1, t, and t^2 are very nearly linearly dependent.

There are two steps which can be taken to improve the situation. The first involves choosing a relative error tolerance which reflects the accuracy of the data and the precision of the arithmetic. If we choose anything between 0.3×10^{-10} and 0.6×10^{-5}, the singular value analysis program which we have outlined will neglect σ_3. The sets of coefficients obtained are, with long precision,

$$c_1 = -0.167 \times 10^{-2}, \qquad c_2 = -0.162 \times 10^1, \qquad c_3 = 0.871 \times 10^{-3},$$

and, with short precision,

$$c_1 = -0.166 \times 10^{-2}, \qquad c_2 = -0.162 \times 10^1, \qquad c_3 = 0.869 \times 10^{-3}.$$

Now they are in much better agreement. Moreover, they are much smaller, which means there will be much less cancellation in evaluation of the quadratic. The predicted populations in 1980 are 212.91 million and 214.96 million. The effect of the short precision is still noticed, but the results are no longer disastrous.

The singular value analysis program also prints out a set of null coefficients associated with the negligible singular value. They are

$$\gamma_1 = -0.9999995, \qquad \gamma_2 = 0.103 \times 10^{-2}, \qquad \gamma_3 = -0.266 \times 10^{-6}.$$

For t between 1900 and 1970, the value of $|\gamma_1 + \gamma_2 t + \gamma_3 t^2|$ does not exceed 0.0017. So for any α the coefficients can be changed from c_j to $c_j + \alpha \gamma_j$ without altering the values produced by the model by more than 0.0017α. Any of the four sets of coefficients we computed could be obtained from any of the others by such a change.

The other step which can be taken to improve the situation is to change the basis. Models of the form

$$y(t) \approx c_1 + c_2(t - 1900) + c_3(t - 1900)^2$$

or

$$y(t) \approx c_1 + c_2 \left(\frac{t - 1935}{10} \right) + c_3 \left(\frac{t - 1935}{10} \right)^2$$

are much more satisfactory. The important thing is to transform the independent variable from the range [1900, 1970] to something more reasonable, like [0, 70], or, better still, [−3.5, 3.5]. The singular values of the design matrices obtained with these two models are

$$\sigma_1 = 0.684 \times 10^4, \qquad \sigma_2 = 0.293 \times 10^2, \qquad \sigma_3 = 0.119 \times 10^1$$

and

$$\sigma_1 = 0.198 \times 10^2, \qquad \sigma_2 = 0.648 \times 10^1, \qquad \sigma_3 = 0.185 \times 10^1,$$

respectively. The condition numbers are 0.575×10^4 and 10.7, which are more than acceptable, even for short-precision computation.

The computations can now be done in either precision and, unless the error tolerance chosen is quite large, none of the singular values will be regarded as negligible. All the fits done in this way give reliable coefficients, and all of them predict the 1980 population as about 227.78 million.

We are still not sure what the population will be in 1980. In an attempt to find out, the degree of the polynomial might be increased. The choice of basis polynomials and error tolerance becomes more critical. It is even possible, and probably reasonable, to use models involving functions other than polynomials. Exploration of these possibilities is left as a problem. The point we wish to make is that the use of the singular value decomposition provides valuable information about the reliability and sensitivity of least-squares models and helps the designer of such models assess their usefulness.

9.2. ORTHOGONALITY AND THE SVD

The singular value decomposition—the SVD—is a powerful computational tool for analyzing matrices and problems involving matrices which has applications in many fields. In the remaining sections of this chapter, we shall define the SVD, describe some other applications, and present an algorithm for computing it. The algorithm is representative of algorithms currently used for various matrix eigenvalue problems and serves as an introduction to computational techniques for these problems as well.

Although it is still not widely known, the singular value decomposition has a fairly long history. Much of the fundamental work was done by Gene Golub and his colleagues W. Kahan, Peter Businger, and Christian Reinsch. Our discussion will be based largely on a paper by Golub and Reinsch (1971). The underlying matrix eigenvalue algorithms have been developed by J. G. F. Francis, H. Rutishauser, and J. H. Wilkinson and are presented in Wilkinson's book (1965). Recent books by Lawson and Hanson (1974) and Stewart (1973) discuss the SVD as well as many related topics.

In elementary linear algebra, a set of vectors is defined to be *independent* if none of them can be expressed as a linear combination of the others. In computational linear algebra, it is very useful to have a quantative notion of the "amount" of independence. We would like to define a quantity that reflects the fact that, for example,

$$\begin{pmatrix} 1 \\ 0 \\ 0 \end{pmatrix}, \begin{pmatrix} 0 \\ 1 \\ 0 \end{pmatrix}, \text{ and } \begin{pmatrix} 0 \\ 0 \\ 1 \end{pmatrix}$$

are *very independent*, whereas

$$\begin{pmatrix} 1.01 \\ 1.00 \\ 1.00 \end{pmatrix}, \quad \begin{pmatrix} 1.00 \\ 1.01 \\ 1.00 \end{pmatrix}, \quad \text{and} \quad \begin{pmatrix} 1.00 \\ 1.00 \\ 1.01 \end{pmatrix}$$

are *almost dependent*.

Since two vectors are dependent if they are parallel, it is reasonable to regard them as very independent if they are perpendicular or orthogonal. Using a superscript T to denote the transpose of a vector or matrix, two vectors u and v are *orthogonal* if their inner product is zero, that is, if

$$u^T v = 0.$$

Moreover, a vector u has length 1 if

$$u^T u = 1.$$

A square matrix is called *orthogonal* if its columns are mutually orthogonal vectors each of length 1. Thus a matrix U is orthogonal if

$$U^T U = I, \quad \text{the identity matrix.}$$

Note that an orthogonal matrix is automatically nonsingular, since $U^{-1} = U^T$. In fact, we shall soon make precise the idea that an orthogonal matrix is *very nonsingular* and that its columns are *very independent*.

The simplest examples of orthogonal matrices are planar rotations of the form

$$U = \begin{pmatrix} \cos\theta & \sin\theta \\ -\sin\theta & \cos\theta \end{pmatrix}.$$

If x is a vector in 2-space, then Ux is the same vector rotated through an angle θ. It is useful to associate orthogonal matrices with such rotations, even though in higher dimensions orthogonal matrices can be more complicated. For example,

$$U = \frac{1}{49} \begin{pmatrix} 24 & 36 & 23 \\ 41 & -12 & -24 \\ 12 & -31 & 36 \end{pmatrix}$$

is orthogonal but cannot be interpreted as a simple plane rotation.

Multiplication by orthogonal matrices does not change such important geometrical quantities as the length of a vector or the angle between two vectors. Orthogonal matrices also have highly desirable computational properties because they do not magnify errors.

For any matrix A and any two orthogonal matrices U and V, consider the matrix Σ defined by

$$\Sigma = U^T A V.$$

If u_j and v_j are the columns of U and V, respectively, then the individual components of Σ are

$$\sigma_{ij} = u_i^T A v_j.$$

The idea behind the singular value decomposition is that by proper choice of U and V it is possible to make most of the σ_{ij} zero; in fact, it is possible to make Σ *diagonal* with nonnegative diagonal entries. Consequently, we make the following definition.

A *singular value decomposition* of an m-by-n real matrix A is any factorization of the form

$$A = U\Sigma V^T,$$

where U is an m-by-m orthogonal matrix, V is an n-by-n orthogonal matrix, and Σ is an m-by-n diagonal matrix with $\sigma_{ij} = 0$ if $i \neq j$ and $\sigma_{ii} = \sigma_i \geq 0$. The quantities σ_i are called the *singular values* of A, and the columns of U and V are called the left and right *singular vectors*.

Readers familiar with matrix eigenvalues should note that the matrices AA^T and A^TA have the same nonzero eigenvalues and that the singular values of A are the positive square roots of these eigenvalues. Moreover, the left and right singular vectors are particular choices of the eigenvectors of AA^T and A^TA, respectively.

In the language of abstract linear algebra, the matrix A is the representation of some linear transformation in a particular coordinate system. By making one orthogonal change of coordinates in the domain of this transformation and a second orthogonal change of coordinates in the range, the representation becomes diagonal.

To simplify notation, we shall assume that all rectangular matrices have at least as many rows as columns, so that $m \geq n$. In the application of the SVD to analysis of experimental data, the matrix element a_{ij} represents the ith observation of the jth variable in an experiment, so m is the total number of observations and n is the total number of variables. We have seen problems with $m = 10,000$ and $n = 5$, but problems with $m = n$ are probably the most common. The assumption $m \geq n$ implies there are n singular values, σ_i, $i = 1, \ldots, n$.

There is some arbitrariness in the SVD. The above definition did not specify any particular order for the singular values. Even if the order was specified, the columns of U and V associated with multiple singular values are not uniquely determined. In actual computation, if the subroutine SVD described later is given the same matrix A on different computers with different word lengths or even just different square root routines, it may produce quite different matrices U and V, but the results on any one machine will satisfy the definition to within the accuracy of that machine and will provide satisfactory solutions to our problems.

Here is a 5-by-3 example which we shall use throughout the chapter. (The entries are given to only three decimal places to save space, even though in practice we would usually work with more figures.)

$$A = \begin{pmatrix} 1 & 6 & 11 \\ 2 & 7 & 12 \\ 3 & 8 & 13 \\ 4 & 9 & 14 \\ 5 & 10 & 15 \end{pmatrix},$$

$$U = \begin{pmatrix} 0.355 & -0.689 & 0.541 & 0.193 & 0.265 \\ 0.399 & -0.376 & -0.802 & -0.113 & 0.210 \\ 0.443 & -0.062 & 0.160 & -0.587 & -0.656 \\ 0.487 & 0.251 & -0.079 & 0.742 & -0.378 \\ 0.531 & 0.564 & 0.180 & -0.235 & 0.559 \end{pmatrix},$$

$$\Sigma = \begin{pmatrix} 35.127 & 0 & 0 \\ 0 & 2.465 & 0 \\ 0 & 0 & 0 \\ 0 & 0 & 0 \\ 0 & 0 & 0 \end{pmatrix},$$

$$V = \begin{pmatrix} 0.202 & 0.890 & 0.408 \\ 0.517 & 0.257 & -0.816 \\ 0.832 & -0.376 & 0.408 \end{pmatrix}.$$

Notice that because of the zeros in Σ, at most the first n columns of U actually contribute to the product $U\Sigma V^T$. Furthermore, if some of the singular values are zero, then fewer than n columns of U are needed. If k is the number of nonzero singular values, then it is possible to make U m by k, Σ k by k, and V^T k by n. Such matrices U and V are technically not orthogonal because they are not square, but their columns form a set of orthonormal vectors. This version of the SVD might be called the *economy size*. The savings in computer storage can be quite important if k or n is much smaller than m. The economy version of our example is

$$U = \begin{pmatrix} 0.355 & -0.689 \\ 0.399 & -0.376 \\ 0.433 & -0.062 \\ 0.487 & 0.251 \\ 0.531 & 0.564 \end{pmatrix},$$

$$\Sigma = \begin{pmatrix} 35.127 & 0 \\ 0 & 2.465 \end{pmatrix},$$

$$V = \begin{pmatrix} 0.202 & 0.890 \\ 0.517 & 0.257 \\ 0.832 & -0.376 \end{pmatrix}.$$

The notion of the *rank* of a matrix is fundamental to much of linear algebra. The usual definition is the maximum number of independent columns, or, equivalently, the order of the maximal nonzero subdeterminant in the matrix. Using such a definition, it is difficult to actually determine the rank of a general matrix in practice. However, if the matrix is diagonal, it is clear that its rank is the number of nonzero diagonal entries. If a set of independent vectors is multiplied by an orthogonal matrix, the resulting set is still independent. In other words, the rank of a general matrix A is equal to the rank of the diagonal matrix Σ in its SVD. Consequently, a practical definition of the rank of a matrix is the number of nonzero singular values. We shall use the letter k to denote rank. (In the example, $k = 2$.)

An m-by-n matrix with $m \geq n$ is said to be of *full rank* if $k = n$ or *rank deficient* if $k < n$. For square matrices, the more common terms *nonsingular* and *singular* are often used for full rank and rank deficient, respectively.

Since the rank of a matrix must always be an integer, it is necessarily a discontinuous function of the elements of the matrix. Arbitrarily small changes (such as roundoff errors) in a rank-deficient matrix can make all of its singular values nonzero and hence create a matrix which is technically of full rank. In practice, we work with the *effective rank*, the number of singular values greater than some prescribed tolerance which reflects the accuracy of the data. This is also a discontinuous function, but the discontinuities are much less numerous and troublesome than those of the theoretical rank.

The great advantage of the use of the SVD in determining the rank of a matrix is that decisions need be made only about the negligibility of single numbers—the small singular values—rather than vectors or sets of vectors.

We can now precisely define the measure of independence mentioned earlier. The *condition number* of a matrix A of full rank is

$$\text{cond}\,(A) = \frac{\sigma_{\max}}{\sigma_{\min}},$$

where $\sigma_{\max}$ and $\sigma_{\min}$ are the largest and smallest singular values of A. If A is rank deficient, then $\sigma_{\min} = 0$ and cond (A) is said to be infinite.

The condition number defined in Chapter 3 and estimated by subroutine DECOMP is based on a different vector norm and so may have a different value than the condition number defined here. The Chapter 3 condition number is easier to compute but does not have as many convenient properties. The two condition numbers will usually have comparable values, however.

Clearly, cond $(A) \geq 1$. If cond (A) is close to 1, then the columns of A are *very independent*. If cond (A) is large, then the columns of A are *nearly dependent*. If A is square, then terms like *nearly singular* or *far from singular* can be given fairly precise meanings. A matrix A is considered to be *more singular* than a matrix B if cond $(A) >$ cond (B).

If A is orthogonal, then cond $(A) = 1$, and so the columns of an orthogonal matrix are as independent as possible. Conversely, if cond $(A) = 1$, then it turns out that A must be a scalar multiple of an orthogonal matrix.

The length or *norm of a vector* is

$$\|x\| = (x^T x)^{1/2}.$$

The quantity $\|Ax\|/\|x\|$ measures the amount by which A *stretches* x. As x ranges over all (nonzero) vectors, how does the stretch factor $\|Ax\|/\|x\|$ vary? Let $A = U\Sigma V^T$ and $y = V^T x$. Then, since orthogonal matrices preserve length, $\|y\| = \|x\|$, and

$$\|Ax\| = \|U\Sigma V^T x\| = \|\Sigma y\|.$$

Since Σ is diagonal, it is clear that

$$\sigma_{\min}\|y\| \leq \|\Sigma y\| \leq \sigma_{\max}\|y\|.$$

Consequently,

$$\sigma_{\min} \leq \frac{\|Ax\|}{\|x\|} \leq \sigma_{\max}.$$

The singular values can be intrepreted geometrically. The matrix A maps the unit sphere, which is the set of vectors x for which $\|x\| = 1$, onto a set of vectors $b = Ax$ which have varying lengths. The image set is actually a k-dimensional ellipsoid embedded in m-dimensional space. The situation where $m = n = k = 2$ is shown in Fig. 9.1. The singular values are the lengths of the various axes of the ellipsoid. The extreme singular values $\sigma_{\max}$ and $\sigma_{\min}$ are the lengths of the major and minor axes. The condition number is related to the eccentricity of the ellipsoid, with large condition numbers corresponding to very eccentric ellipsoids.

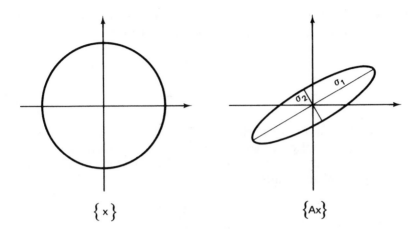

$\{x\}$ $\{Ax\}$

Fig. 9.1.

The *norm of a matrix* can be defined to be the maximum stretch factor, which is simply'

$$\|A\| = \sigma_{max}.$$

The reader may recall that we used a different definition for $\|A\|$ in Chapter 3. The Chapter 3 definition is much less expensive to evaluate than σ_{max}, but it is not so well suited to the types of problems we are dealing with in this chapter.

9.3. APPLICATIONS

In this section, we shall present a few other applications of the singular value decomposition.

The General Linear Equations Problem

Let A be a given m-by-n matrix where $m \geq n$, and let b be a given m vector. Find all n vectors x which solve

$$Ax = b.$$

Notice that the important case where A is square but possibly singular is included. This problem involves such questions as, Are the equations consistent? Do any solutions exist? Is the solution unique? Does $Ax = 0$ have any nonzero solutions? What is the general form of the solution?

Theoretically, there are many different algorithms which answer these questions. But in computational practice with inexact data and imprecise arithmetic, the SVD is essentially the only known method which is reliable.

Using the SVD of A, the linear system $Ax = b$ becomes

$$U\Sigma V^T x = b,$$

and hence

$$\Sigma z = d,$$

where $z = V^T x$ and $d = U^T b$. The system of equations $\Sigma z = d$ is diagonal and hence can be easily studied. It breaks up into as many as three sets, depending on the values of the dimensions m and n and the rank k, the number of nonzero singular values:

$$\sigma_j z_j = d_j, \quad \text{if } j \leq n \text{ and } \sigma_j \neq 0,$$
$$0 \cdot z_j = d_j, \quad \text{if } j \leq n \text{ and } \sigma_j = 0,$$
$$0 = d_j, \quad \text{if } j > n.$$

The second set of equations is empty if $k = n$, and the third is empty if $n = m$.

The equations are consistent and a solution exists if and only if $d_j = 0$ whenever $\sigma_j = 0$ or $j > n$. If $k < n$, then the z_j associated with a zero σ_j can be given an arbitrary value and still yield a solution to the system. When transformed back to the original coordinates by $x = Vz$, these arbitrary components of z serve to parameterize the space of all possible solutions x.

Let u_j and v_j be the columns of U and V. Then the decomposition $A = U\Sigma V^T$ can be written

$$Av_j = \sigma_j u_j, \quad j = 1, \ldots, n.$$

(Because we put a superscript T on V in the basic SVD equation, this equation now involves the columns of both U and V, rather than the rows of one and the columns of the other.) The *null space* of A is the set of x for which $Ax = 0$, and the *range* of A is the set of b for which $Ax = b$ has a solution x. If $\sigma_j = 0$, then $Av_j = 0$ and v_j is in the null space of A, while if $\sigma_j \neq 0$, then u_j is in the range of A.

Consequently, we can get a complete description of the null space and range in the following way. Let V_0 be the set of columns v_j for which $\sigma_j = 0$, and let V_1 be the remaining columns v_j. Let U_1 be the set of columns u_j for which $\sigma_j \neq 0$, and let U_0 be the remaining columns u_j, including those with $j > n$. There are k columns in V_0, there are $n - k$ columns in V_1 and in U_1, and there are $m - n + k$ columns in U_0. Furthermore,

1. V_0 is an orthonormal basis for the null space of A.
2. V_1 is an orthonormal basis for the orthogonal complement of the null space of A.

3. U_1 is an orthonormal basis for the range of A.

4. U_0 is an orthonormal basis for the orthogonal complement of the range of A.

The matrix

$$A = \begin{pmatrix} 1 & 6 & 11 \\ 2 & 7 & 12 \\ 3 & 8 & 13 \\ 4 & 9 & 14 \\ 5 & 10 & 15 \end{pmatrix}$$

given earlier is rank deficient. Its middle column is the average of the other two columns. This is reflected by the fact that the last column of V, which is $(0.408, -0.816, 0.408)^T$, has components which are in the ratio $1: -2: 1$. The equations $Ax = b$ will have a solution if and only if b is a linear combination of the first two columns of U. If a solution exists, an arbitrary multiple of the last column of V can be added to it.

For example, let

$$b = \begin{pmatrix} 5 \\ 5 \\ 5 \\ 5 \\ 5 \end{pmatrix}.$$

Then

$$d = U^T b = \begin{pmatrix} 11.071 \\ -1.561 \\ 0 \\ 0 \\ 0 \end{pmatrix}.$$

The last three components of d are zero; hence b was a linear combination of the first two columns of U, and the equations $Ax = b$, as well as $\Sigma z = d$, are consistent. The latter set is

$$35.127z_1 = 11.071,$$
$$2.465z_2 = -1.561,$$
$$0z_3 = 0.$$

The solution is

$$z_1 = 0.315$$
$$z_2 = -0.633$$
$$z_3 = \text{anything.}$$

Taking $z_3 = 0$ and computing $x = Vz$, we obtain

$$x = \begin{pmatrix} -0.500 \\ 0 \\ 0.500 \end{pmatrix}.$$

Observe that $x = (-\frac{1}{2}, 0, \frac{1}{2})^T$ is indeed a solution to $Ax = b$. On the other hand, taking $z_3 = 1.225$ leads to

$$x = \begin{pmatrix} -1.001 \\ 1.002 \\ -0.001 \end{pmatrix},$$

and $x = (-1, 1, 0)^T$ is also a solution. In fact, using the last column of V, the general solution is

$$x = \begin{pmatrix} -0.500 \\ 0 \\ 0.500 \end{pmatrix} + z_3 \begin{pmatrix} 0.408 \\ -0.816 \\ 0.408 \end{pmatrix}$$

$$= \frac{1}{2} \begin{pmatrix} -1 \\ 0 \\ 1 \end{pmatrix} + \alpha \begin{pmatrix} 1 \\ -2 \\ 1 \end{pmatrix}.$$

The choice $z_3 = 0$ has a special significance. It leads to the solution $(-\frac{1}{2}, 0, \frac{1}{2})^T$ which is the shortest of all possible solutions.

For another example, let

$$b = \begin{pmatrix} 4 \\ 5 \\ 5 \\ 5 \\ 5 \end{pmatrix}.$$

Then

$$d = U^T b = \begin{pmatrix} 10.716 \\ -0.872 \\ -0.541 \\ -0.193 \\ -0.265 \end{pmatrix}.$$

Since the last three components of d are not zero, the equation $Ax = b$ has no solutions. In other words b is not in the range of A.

The Linear Least-Squares Problem

This is an extension of the previous problem, but now we seek n vectors x for which Ax is only approximately equal to b in the sense that the length of the residual is minimized. By the residual we mean the m vector

$$r = Ax - b.$$

The problem then is to pick an x which minimizes $||r||^2 = \sum_{i=1}^{m} r_i^2$. (Minimizing the square of the length is equivalent to minimizing the length itself.)

Statisticians refer to this as the *linear regression* problem. The matrix A is usually denoted by X and is called the *design* matrix. The right-hand side b is usually denoted by y and is the *data* vector. Of course, this is the same problem we considered in Section 9.1. Here we are just using more matrix terminology.

If A has full rank, then the solution x is unique and can be reliably computed by several different algorithms, some of which are faster than the SVD. But the SVD also handles the rank-deficient case and, except for some very large problems, is not much more costly than the other reliable methods. (It is less costly than a fast algorithm which may give the wrong answer.)

Since orthogonal matrices preserve norm,

$$||r|| = ||U^T(AVV^Tx - b)|| = ||\Sigma z - d||.$$

Consequently, the SVD reduces the general least-squares problem to one involving a diagnoal matrix. It is easy to see that the vector z that produces the minimum $||r||$ is given by

$$z_j = \frac{d_j}{\sigma_j}, \qquad \text{if } \sigma_j \neq 0.$$

$$z_j = \text{anything}, \qquad \text{if } \sigma_j = 0.$$

So k of the equations in diagonal form are solved exactly. The remaining ones lead to a possibly nonzero residual vector with norm given by $||r||^2 = \sum d_j^2$, where the sum is over all j for which $\sigma_j = 0$, together with $j > n$. The back transformation $x = Vz$ then solves the original problem.

If the problem is rank deficient, then the solution which minimizes $||r||$ is not unique. In this situation it is frequently desirable to obtain uniqueness by picking the shortest such solution. This is obtained by setting

$$z_j = 0, \qquad \text{if } \sigma_j = 0.$$

In the full-rank case there is a unique solution.

Sometimes least-squares problems are modified by considering the minimization of some combination of $\|r\|$ and $\|x\|$. The solutions of such problems are obtained easily from the SVD by methods which are described by Lawson and Hanson (1974).

As an example, consider again the problem with

$$b = \begin{pmatrix} 4 \\ 5 \\ 5 \\ 5 \\ 5 \end{pmatrix}.$$

As we saw, this is an inconsistent right-hand side, and so $Ax = b$ cannot be solved exactly. The diagonalized equations are

$$35.127z_1 \approx 10.716,$$
$$2.465z_2 \approx -0.872,$$
$$0z_3 \approx -0.541,$$
$$0 \approx -0.193,$$
$$0 \approx -0.265.$$

The symbol $\approx$ means these equations are to be solved in the least-squares sense. The first two can be solved exactly by taking

$$z_1 = 0.305,$$
$$z_2 = -0.354.$$

The third equation does not determine z_3, so we take

$$z_3 = 0$$

to obtain the shortest possible solution. The last three equations tell us that $\|r\|^2$ will be $0.541^2 + 0.193^2 + 0.265^2 = 0.400$. Notice that this can be computed without actually finding x, although it is somewhat more accurate to compute x and then $\|b - Ax\|$, especially if the residuals are near zero.

The solution $x = Vz$ is

$$x = \begin{pmatrix} -0.253 \\ 0.067 \\ 0.387 \end{pmatrix}.$$

This leads to

$$Ax = \begin{pmatrix} 4.406 \\ 4.607 \\ 4.808 \\ 5.009 \\ 5.210 \end{pmatrix},$$

which is reasonably close to the given b. Again, any multiple of the last column of V can be added to x to produce other, but longer, solutions to the least-squares problem.

The Pseudoinverse

An n-by-m matrix X is called the Moore-Penrose pseudoinverse of A if it satisfies the following four conditions:

1. $AXA = A$.
2. $XAX = X$.
3. AX is symmetric.
4. XA is symmetric.

It can be shown that such an X always exists and is unique. If A is square and nonsingular, then $X = A^{-1}$ obviously satisfies these conditions. If A is rectangular and of full rank, then $X = (A^T A)^{-1} A^T$. If A is not of full rank, there is no such simple representation. The notation A^t is often used for the pseudoinverse.

To use the SVD to obtain the pseudoinverse of a matrix, we begin with the 1-by-1 case, that is, the pseudoinverse of a number:

$$\sigma^t = \begin{cases} \dfrac{1}{\sigma}, & \text{if } \sigma \neq 0, \\ 0, & \text{if } \sigma = 0. \end{cases}$$

Note that $\sigma \cdot \sigma^t = 0$ or 1 and that the four Moore-Penrose conditions are clearly satisfied.

The pseudoinverse of an m-by-n diagonal matrix

$$\Sigma = \begin{pmatrix} \sigma_1 & & & & \\ & \sigma_2 & & 0 & \\ & & \cdot & & \\ & 0 & & \cdot & \\ & & & & \sigma_n \\ & & 0 & & \end{pmatrix}$$

can then be defined to be the n-by-m diagonal matrix

$$\Sigma^\dagger = \begin{pmatrix} \sigma_1^\dagger & & & \\ & \sigma_2^\dagger & 0 & \\ & & \ddots & \\ & 0 & & \ddots & 0 \\ & & & & \sigma_n^\dagger \end{pmatrix}.$$

Note that $\Sigma\Sigma^\dagger = I_k$, a matrix with k 1's on the diagonal and 0's elsewhere.

Finally, using a three-term analog of the familiar identity $(AB)^{-1} = B^{-1}A^{-1}$, the pseudoinverse of a general matrix can be obtained from its SVD by

$$A^\dagger = V\Sigma^\dagger U^T.$$

It is easy to verify that this definition satisfies the four Moore-Penrose equations.

If A is square and nonsingular, then all its singular values are nonzero, and hence $\Sigma^\dagger$ is Σ^{-1} and $A^\dagger$ is A^{-1}.

The pseudoinverse is related to the least-squares problem by the fact that the shortest x which minimizes $\|Ax - b\|$ can be expressed as $x = A^\dagger b$.

If the basic relation

$$A = U\Sigma V^T$$

is rewritten in terms of individual elements, we find for $i = 1, \ldots, m$ and $j = 1, \ldots, n$ that

$$a_{ij} = \sum_{k=1}^{n} \sigma_k u_{ik} v_{jk}.$$

Of course, the sum could be restricted to those values of k for which $\sigma_k \neq 0$. If the definition

$$A^\dagger = V\Sigma^\dagger U^T$$

is also rewritten, then for $i = 1, \ldots, n$ and $j = 1, \ldots, m$

$$a_{ij}^\dagger = \sum_{\sigma_k \neq 0} \frac{v_{ik} u_{jk}}{\sigma_k}.$$

Now it is essential to restrict the sum to nonzero σ_k.

It should be emphasized that the elements of the pseudoinverse are not continuous functions of the elements of the matrix. There are *jumps* in the pseudoinverse of a varying matrix whenever the matrix changes rank. Consequently, in computational practice it is important to compare the small singular values with some tolerance that reflects the errors in the data and computation. An *effective pseudoinverse* can be defined in terms of such a tolerance τ as follows:

$$\sigma_\tau^\dagger = \begin{cases} \dfrac{1}{\sigma}, & \text{if } \sigma > \tau, \\[2mm] 0, & \text{otherwise.} \end{cases}$$

This leads to an $A_\tau^\dagger$, which has elements

$$a_{ij}^\dagger = \sum_{\sigma_k > \tau} \frac{v_{ik} u_{jk}}{\sigma_k}.$$

Only the last three of the Moore-Penrose conditions are still satisfied by $X = A_\tau^\dagger$, that is,

$$XAX = X;$$

$$AX \text{ is symmetric};$$

$$XA \text{ is symmetric}.$$

The first one is replaced by an inequality involving the matrix norm defined earlier:

$$\| AXA - A \| < \tau.$$

If $\tau \neq 0$, the matrix X which satisfies these four conditions may not be unique, but $X = A_\tau^\dagger$ has the additional property that it minimizes $\| X \|$.

It is this ability to handle inexact data and imprecise computation that makes the SVD characterization of the pseudoinverse preferable to other approaches.

Again, referring to our example, we find that

$$A^\dagger = \begin{pmatrix} -0.247 & -0.133 & -0.020 & 0.093 & 0.207 \\ -0.067 & -0.033 & 0.000 & 0.033 & 0.067 \\ 0.113 & 0.067 & 0.020 & -0.027 & -0.073 \end{pmatrix}.$$

In this example $A^\dagger A$ is nowhere near the identity.

Approximating Matrices

If A is a moderately large matrix, say with m and n on the order of a few hundred, then storing all the elements of A in a computer may require a significant portion of the total memory available. It may be desirable to approximate A with a "simpler" matrix which requires less storage. A reasonable meaning of a *simple* matrix is one of relatively low rank, that is, with only a few independent columns. In addition to saving storage, such matrices may provide important insight into particular problems.

The equation $A = U\Sigma V^T$ can also be written

$$A = \sigma_1 E_1 + \sigma_2 E_2 + \ldots + \sigma_n E_n,$$

where $E_i = u_i v_i^T$ is the *outer product* of a column of U with the corresponding column of V. Each E_i is a simple matrix of rank 1, which can be stored using only $m + n$ locations rather than mn locations. Furthermore, the product $E_i x$ can be formed for any vector x with only $m + n$ multiplications rather than mn. To appreciate the magnitude of these savings, note that when $m = n = 300$, then $m + n = 600$, while $mn = 90,000$.

The sum of the squares of all the elements in an E_i is 1, so none of the E_i are numerically more or less important than any of the others. However, some of the coefficients σ_i may well be significantly smaller than some of the others, and so the above series of A can be truncated after a few terms. The truncated sum defines a matrix which approximates A but which can be stored and manipulated with less cost. Of course, the usefulness of this procedure depends on the particular A and the distribution of its singular values.

Such approximations have recently been used to advantage in such diverse fields as digital image processing and cryptography; see Andrews and Patterson (1975) and Moler and Morrison (1977).

Evaluating Determinants and Finding Singular Matrices

Let $A(z)$ be a square matrix which depends in a possibly complicated way on some parameter z. Consider finding a value, or values, of z for which $A(z)$ is singular. Equivalently, find z for which the determinant of $A(z)$ is zero.

Using det to denote determinant,

$$\det(A) = \det(U) \cdot \det(\Sigma) \cdot \det(V^T).$$

The determinant of an orthogonal matrix is ± 1, and the determinant of a diagonal matrix is simply the product of its elements. So

$$\det(A) = \pm\sigma_1 \cdot \sigma_2 \cdot \ldots \cdot \sigma_n.$$

Computing determinants can be tricky because the value can vary over a huge range, and floating-point underflows and overflows are frequent.

Many problems which are formulated in terms of determinants do not require the actual value of the determinant but simply some indication of when it is zero. It is theoretically possible to compute $|\det(A)|$ by taking the product of the singular values of A. But in most situations, it is sufficient to use only the smallest singular value and thereby avoid underflow/overflow and other numerical difficulties.

A general-purpose minimizer, such as FMIN, can be used to vary z in a systematic search for a value which makes $\sigma_{\min}(A(z))$ as small as possible, hopefully zero. We have had considerable success with this technique in a fairly wide range of problems.

In principle, it is possible to modify the SVD computation to return $+1$ or -1 for det $(U) \cdot \det(V^T)$. It would then be possible to compute the sign of the determinant and to use a zero finder, such as ZEROIN, rather than a minimizer. We have not yet had much experience with this idea.

The Eigenvalue Problem

This is not really a practical application of SVD but rather an indication of how the SVD is related to a larger class of important matrix problems.

Let A be a square matrix. The eigenvalues of A are the numbers λ for which

$$Ax = \lambda x$$

has a nonzero solution x. Since this is equivalent to requiring

$$\det(A - \lambda I) = 0$$

the eigenvalues could theoretically be computed by finding the roots of this polynomial. However, consider

$$A = \begin{pmatrix} 1 & e \\ e & 1 \end{pmatrix}$$

for some small but not negligible number e. The true eigenvalues are $\lambda_1 = 1 + e$ and $\lambda_2 = 1 - e$. The polynomial $\det(A - \lambda I)$ is

$$\lambda^2 - 2\lambda + (1 - e^2).$$

If e is small, then e^2 is much smaller, and it is necessary to have *twice* the precision in the coefficients of the polynomial than is present in either the matrix elements or the eigenvalues. This difficulty becomes even more pronounced for higher-order matrices. Consequently, modern methods for computing eigenvalues avoid the use of polynomials or polynomial root finders.

The connection between SVD and eigenvalues is simplest for matrices which are symmetric and positive semidefinite. Symmetry is easy to define and check—A is symmetric if $a_{ij} = a_{ji}$ for all i and j. Positive semidefiniteness is much more elusive—positive semidefinite means that $x^T A x \geq 0$ for all x. Roughly this means that the diagonal elements of A are fairly large compared to the off-diagonal elements. In fact, a sufficient condition is that for each i

$$a_{ii} \geq \sum_{j \neq 1} |a_{ij}|,$$

but this is not necessary.

It is not difficult to show that if A is real and symmetric, then all its eigenvalues are real, and if A is positive semidefinite, then all its eigenvalues are nonnegative. If $A = U\Sigma V^T$ and $A = A^T$, then

$$A^2 = A^T A = V\Sigma^T U^T U \Sigma V^T$$
$$= V\Sigma^2 V^T.$$

Thus,

$$A^2 V = V\Sigma^2.$$

Letting v_j denote the columns of V and looking at the jth column of this equation, we find

$$A^2 v_j = \sigma_j^2 v_j.$$

Since the eigenvalues of A^2 are the squares of the eigenvalues of A and since the eigenvalues of A are nonnegative, we conclude that the eigenvalues of a symmetric, positive-semidefinite matrix are equal to its singular values and that the eigenvectors are the columns of V.

If A is symmetric but not positive semidefinite, then some of its eigenvalues are negative. It turns out that the absolute values of the eigenvalues are equal to the singular values and that the signs of the eigenvalues can be recovered by comparing the columns of U with those of V, but we shall not go into details. If A is not symmetric, there is no simple connection between its eigenvalues and its singular values.

This shows that the SVD algorithm could be used to compute the eigenvalues of symmetric matrices. However, this is not particularly efficient and not recommended. Our only point is that the SVD is closely related to matrix eigenvalue problems. Careful study of the SVD algorithm will help in understanding eigenvalue algorithms as well.

9.4. COMPUTING THE DECOMPOSITION

SVD is also the name of a computer subroutine for computing the singular value decomposition of an arbitrary matrix. It is based upon an Algol procedure by Golub and Reinsch (1971).

Before describing how SVD works, it is useful to review a more elementary matrix algorithm—Gaussian elimination—from a slightly different viewpoint.

Let

$$A_1 = \begin{pmatrix} 1 & 6 & 11 \\ 2 & 7 & 12 \\ 3 & 8 & 13 \\ 4 & 9 & 14 \\ 5 & 10 & 15 \end{pmatrix}.$$

If A_1 were the coefficient matrix in a set of linear equations, then the first step in solving the system would be to subtract multiples of the first row from the other rows in order to produce zeros in the first column. (It is not necessary to pivot in this simple example.) We would obtain

$$A_2 = \begin{pmatrix} 1 & 6 & 11 \\ 0 & -5 & -10 \\ 0 & -10 & -20 \\ 0 & -15 & -30 \\ 0 & -20 & -40 \end{pmatrix}.$$

How can the transformation of A_1 into A_2 be described using matrix notation? Let

$$m_1 = \begin{pmatrix} 0 \\ 2 \\ 3 \\ 4 \\ 5 \end{pmatrix}$$

$$e_1^T = (1 \ \ 0 \ \ 0 \ \ 0 \ \ 0).$$

Using the outer product $m_1 e_1^T$, form the matrix $M_1 = I - m_1 e_1^T$:

$$M_1 = \begin{pmatrix} 1 & 0 & 0 & 0 & 0 \\ -2 & 1 & 0 & 0 & 0 \\ -3 & 0 & 1 & 0 & 0 \\ -4 & 0 & 0 & 1 & 0 \\ -5 & 0 & 0 & 0 & 1 \end{pmatrix}.$$

It is easy to check that

$$A_2 = M_1 A_1.$$

Of course, a subroutine for Gaussian elimination does not actually form the matrix M_1 or carry out the matrix multiplication M_1A_1. It simply operates on the rows of A_1 to produce A_2. However, the most compact mathematical description of the process is in terms of M_1. Furthermore, use of M_1 allows the transformation to be viewed as an operation on the columns of A_1 instead of the rows. Let a_j be the jth column of A. Then

$$\begin{aligned} M_1 a_j &= (I - m_1 e_1^T)a_j \\ &= a_j - (m_1 e_1^T)a_j \\ &= a_j - m_1(e_1^T a_j) \\ &= a_j - (a_{1j})m_1. \end{aligned}$$

In particular,

$$M_1 a_1 = a_1 - 1 \cdot m_1 = \begin{pmatrix} 1 \\ 0 \\ 0 \\ 0 \\ 0 \end{pmatrix},$$

$$M_1 a_2 = a_2 - 6 \cdot m_1 = \begin{pmatrix} 6 \\ -5 \\ -10 \\ -15 \\ -20 \end{pmatrix},$$

$$M_1 a_3 = a_3 - 11 \cdot m_1 = \begin{pmatrix} 11 \\ -10 \\ -20 \\ -30 \\ -40 \end{pmatrix}.$$

This shows that the columns of A_2 can be obtained by subtracting certain multiples of m_1 from the columns of A_1.

The next step of the elimination would produce

$$A_3 = \begin{pmatrix} 1 & 6 & 11 \\ 0 & -5 & -10 \\ 0 & 0 & 0 \\ 0 & 0 & 0 \\ 0 & 0 & 0 \end{pmatrix}.$$

The step can be described using

$$m_2 = \begin{pmatrix} 0 \\ 0 \\ 2 \\ 3 \\ 4 \end{pmatrix}.$$

$$e_2^T = (0 \quad 1 \quad 0 \quad 0 \quad 0)$$

and

$$M_2 = I - m_2 e_2^T,$$
$$A_3 = M_2 A_2.$$

The SVD algorithm involves analogous, but somewhat more complicated, ideas. The actual subroutine involves various operations on both the rows and columns of the matrix. The clearest description of the algorithm uses orthogonal analogs of the matrices M_1 and M_2.

The objective of the algorithm is to find orthogonal U and V so that

$$U^T A V = \Sigma \text{ is diagonal.}$$

Both U and V are obtained as the product of orthogonal matrices known as Householder reflections.

There are two stages in the algorithm. The first stage involves the reduction of A to *bidiagonal* form, which is a matrix whose only nonzero elements are on the diagonal and the first superdiagonal. The second stage is an iterative process in which the superdiagonal elements are reduced to a negligible size, leaving the desired diagonal matrix.

Again, consider introducing zeros into a_1, the first column of A_1. Gaussian elimination cannot be used because M_1 is not orthogonal. Instead, let

$$u_1 = \begin{pmatrix} 8.416 \\ 2.000 \\ 3.000 \\ 4.000 \\ 5.000 \end{pmatrix}.$$

This is obtained from a_1 by adding $\|a_1\|$ to the first component of a_1. Let

$$\beta_1 = \frac{\|u_1\|^2}{2} = 62.415,$$
$$U_1 = I - \beta_1^{-1} u_1 u_1^T,$$
$$A_2 = U_1 A_1.$$

The transformed matrix A_2 is actually computed column by column:

$$U_1 a_1 = (I - \beta_1^{-1} u_1 u_1^T) a_1$$
$$= a_1 - \beta_1^{-1} u_1 u_1^T a_1$$
$$= a_1 - u_1 \qquad \text{since } \beta_1^{-1} u_1^T a_1 = 1$$
$$= \begin{pmatrix} -7.416 \\ 0 \\ 0 \\ 0 \\ 0 \end{pmatrix}.$$

Notice that the desired zeros have been introduced into the first column and that its length is unchanged:

$$U_1 a_2 = a_2 - \beta_1^{-1} u_1 u_1^T a_2$$
$$= a_2 - (2.796) u_1$$
$$= \begin{pmatrix} -17.529 \\ 1.409 \\ -0.387 \\ -2.183 \\ -3.949 \end{pmatrix},$$

$$U_1 a_3 = a_3 - \beta_1^{-1} u_1 u_1^T a_3$$
$$= a_3 - (4.592) u_1$$
$$= \begin{pmatrix} -27.642 \\ 2.817 \\ -0.774 \\ -4.366 \\ -7.957 \end{pmatrix}.$$

Consequently,

$$A_2 = \begin{pmatrix} -7.416 & -17.529 & -27.642 \\ 0 & 1.409 & 2.817 \\ 0 & -0.387 & -0.774 \\ 0 & -2.183 & -4.366 \\ 0 & -3.979 & -7.957 \end{pmatrix}.$$

The columns of A_2 are obtained by subtracting multiples of u_1 from the columns of A_1. The multiples come from inner products, rather than from single elements as in Gaussian elimination.

The transforming matrix U_1 is a Householder reflection. Its orthogonality, which is crucial, can be verified by checking that $U_1^T U_1 = I$:

$$
\begin{aligned}
U_1^T U_1 &= (I - \beta_1^{-1} u_1 u_1^T)^T (I - \beta_1^{-1} u_1 u_1^T) \\
&= I - 2\beta_1^{-1} u_1 u_1^T + \beta_1^{-2} u_1 u_1^T u_1 u_1^T \\
&= I - 2\beta_1^{-1} u_1 u_1^T + 2\beta_1^{-1} u_1 u_1^T \\
&= I.
\end{aligned}
$$

This orthogonality guarantees that the length of each column of A_2 is equal to the length of the corresponding column of A_1.

Geometrically, the transformation of any column a_j into $U_1 a_j$ is obtained by reflecting a_j in a plane through the origin perpendicular to u_1. This plane is oriented so that a_1 is reflected into a vector with four zero components.

Before introducing any more zeros below the diagonal, a transformation of the form

$$ A_3 = A_2 V_1 $$

introduces zeros into the first row. The zeros already in the first column must not be disturbed, and the length of the first row must be preserved, so in this example, where n is only 3, it is possible to introduce only one zero. For a larger matrix, $n - 2$ zeros would be introduced in this step.

The transformation is derived from the first row of A_2:

$$
\begin{aligned}
v_1^T &= (0 \quad -50.261 \quad -27.642), \\
\gamma_1 &= \frac{\|v_1\|^2}{2}, \\
V_1 &= I - \gamma_1^{-1} v_1 v_1^T, \\
A_3 &= A_2 V_1.
\end{aligned}
$$

The ith row of A_3 is obtained by

$$
\begin{aligned}
a_i^T V_1 &= a_i^T (I - \gamma_1^{-1} v_1 v_1^T) \\
&= a_i^T - (\gamma_1^{-1} a_i^T v_1) v_1^T.
\end{aligned}
$$

That is, appropriate multiples of v_1^T are subtracted from each row of A_2. This gives

$$A_3 = \begin{pmatrix} -7.416 & 32.732 & 0 \\ 0 & -3.133 & 0.319 \\ 0 & 0.861 & -0.088 \\ 0 & 4.856 & -0.495 \\ 0 & 8.850 & -0.901 \end{pmatrix}.$$

Since the first component of v_1^T is zero, the zeros in the first column of A_2 remain in A_3. Since V_1 is orthogonal, the length of each row of A_3 is equal to the length of the corresponding row of A_2.

More zeros can now be produced below the diagonal without disturbing those already obtained:

$$A_4 = U_2 A_3 = \begin{pmatrix} -7.416 & 32.732 & 0 \\ 0 & 10.605 & -1.080 \\ 0 & 0 & 0.000 \\ 0 & 0 & 0.000 \\ 0 & 0 & 0.000 \end{pmatrix}.$$

Because this matrix is only rank 2, the third column now happens to have elements on the order of roundoff error on and below the diagonal. These elements are dentoed by 0.000 to distinguish them from exactly zero elements. If the matrix were of full rank, one more transformation would have to be done to introduce zeros in the third column:

$$A_5 = U_3 A_4 = \begin{pmatrix} -7.416 & 32.732 & 0 \\ 0 & 10.605 & 1.080 \\ 0 & 0 & 0.000 \\ 0 & 0 & 0 \\ 0 & 0 & 0 \end{pmatrix}.$$

In this example, the third diagonal element would be exactly zero if there were no roundoff error. The elements below the diagonal in all columns are shown as exact zeros because the transformations were designed to produce zeros there. The last transformation, U_3, could have been the identity in this example, but that is not a Householder reflection. The U_3 actually used also changed the sign of the -1.080 in A_4.

This is the desired bidiagonal matrix, and the first stage is complete. It is not possible to introduce any more zeros directly using orthogonal transformations. For general matrices, n U's and $n - 2$ V's are required to reach this point. The number of transformations does not depend on the row dimension m, but the amount of work to apply each transformation does.

The second stage of the SVD algorithm is an iterative process. Each step reduces the size of the unwanted superdiagonal elements. Eventually these elements become comparable with rounding errors and can safely be neglected. The actual number of steps required depends on the particular matrix and the accuracy of the computer, but the rate of convergence is very rapid, and so the amount of time required by the second stage is usually less than that required by the first. This is particularly true if m is much larger than n since this second stage retains all the zeros introduced below the first n rows.

The second stage is a variant of the QR algorithm, which was introduced by Francis around 1960. It has become one of the most reliable and widely used numerical algorithms. A comprehensive discussion of the details of the algorithm, and in particular a satisfactory explanation of why it works, is beyond the scope of this book. We shall only outline the algorithm's behavior in this one example. The interested reader is advised to consult our references for more details.

Each iterative step begins with a bidiagonal matrix and transforms it into another bidiagonal matrix with smaller off-diagonal elements. We shall examine one such step in our example in detail. Since the last two rows are zero and remain zero throughout, they can be omitted. In addition, we shall show five decimal places in the matrix elements instead of the three we have been using until now. We start with

$$
\begin{array}{ccc}
-7.41620 & 32.73169 & 0 \\
0 & 10.60517 & 1.08016 \cdot \\
0 & 0 & 0.00000
\end{array}
$$

Using the three elements in the bottom right-hand corner, a certain transformation V is calculated, so that when A is replaced by AV, the first two columns are changed:

$$
\begin{array}{ccc}
33.56105 & 0.13897 & 0 \\
10.35262 & -2.30062 & 1.08016 \\
0 & 0 & 0.00000.
\end{array}
$$

Notice than a nonzero element, 10.35262, has been introduced below the diagonal. To get rid of this unwanted element, a transformation U is calculated so that when A is replaced with UA the first two rows are changed:

$$
\begin{array}{ccc}
35.12152 & -0.54535 & 0.31839 \\
0 & -2.23937 & 1.03216 \\
0 & 0 & 0.00000.
\end{array}
$$

The desired zero has returned, but a new nonzero element, 0.31839, has appeared above the superdiagonal. Another V works on the last two columns to eliminate the unwanted element:

$$
\begin{array}{ccc}
35.12152 & 0.63149 & 0 \\
0 & 2.45431 & 0.23771 \\
0 & 0.00000 & 0.00000.
\end{array}
$$

If the bottom right-hand element had not been 0.00000, we would now have another nonzero below the diagonal where the other 0.00000 is. Another U, which in this example is nearly equal to the identity matrix, works with the last two rows:

$$
\begin{array}{ccc}
35.12152 & 0.63149 & 0 \\
0 & 2.45431 & 0.23771 \\
0 & 0 & 0.00000.
\end{array}
$$

We are now back to a bidiagonal matrix. Because of the initial V transformation, both superdiagonal elements have become smaller. (For other matrices, some of the superdiagonal elements may actually get larger on any one step, but eventually they all get smaller.)

The process is repeated. The next bidiagonal matrix obtained is

$$
\begin{array}{ccc}
35.12722 & 0.00309 & 0 \\
0 & 2.46540 & -0.00014 \\
0 & 0 & 0.00000.
\end{array}
$$

Notice that the diagonal elements have changed slightly and that, again, both superdiagonal elements have become smaller.

One more step produces

$$
\begin{array}{ccc}
35.12722 & 0.00002 & 0 \\
0 & 2.46540 & 0.00000 \\
0 & 0 & 0.00000.
\end{array}
$$

The changes in the diagonal elements have occurred in the sixth or seventh decimal place, beyond those we are showing. The last superdiagonal element is now on the order of roundoff error and can be regarded as zero.

One final step reduces the other superdiagonal elements to a negligible size:

$$
\begin{array}{ccc}
35.12722 & 0 & 0 \\
0 & 2.46540 & 0 \\
0 & 0 & 0.00000.
\end{array}
$$

The three diagonal elements are the three singular values.

So far we have not explicitly shown any of the transformations that were used but only their effect on A, because the process is carried out using operations on rows and columns of A rather than actual matrix multiplications. If only the singular values are to be computed, then A is the only two-dimensional array involved in the computation. (In fact, in the subroutine SVD, the bidiagonal matrix is represented by two one-dimensional arrays, one of which eventually returns the singular values.)

To calculate the singular vectors U and V as well as the singular values, two other arrays are initialized to identity matrices, and then each time an operation is carried out on the rows or columns of A the same operation is done to one of the two other arrays. When A eventually reaches diagonal form, the two arrays contain U and V.

This algorithm is typical of those which are used in modern computer subroutines for various matrix eigenvalue problems. Unless the problem is already in a special form, there is an initial reduction of the matrix to one containing many zeros. Then an iterative process is applied to the reduced matrix to produce even more zeros. Orthogonal transformations are usually used throughout.

The initial reduction not only ensures that the iterative process is efficient; it also enhances its convergence properties. For the SVD algorithm, for example, the reduced matrix is bidiagonal, the iterative process is designed to annihilate superdiagonal elements, and a theorem of Wilkinson says that the iteration is guaranteed to converge and that the rate of convergence is very rapid. Wilkinson's theorem is discussed as it applies to the SVD in Lawson and Hanson (1974).

9.5. SUBROUTINE SVD

The Algol procedure for computing the SVD written by Golub and Reinsch has been published in the collection of matrix algorithms edited by Wilkinson and Reinsch (1971). Fortran subroutines for a variety of matrix eigenvalue problems, including a translation of the SVD procedure, have been produced by the NATS Project at Argonne National Laboratory. NATS stands for National Activity to Test Software, and the collection of eigenvalue subroutines is known as EISPACK. A user's guide to part of the second release of EISPACK has been published by Smith et al. (1976). Although SVD is included in the second release, it is not described in the 1976 edition of the user's guide. It will be described in a later edition.

Our version of SVD differs from the EISPACK version in the test for negligible elements. The EISPACK version requires a data statement containing machine epsilon. We eliminated the machine epsilon and replaced the test for negligible elements with statements like

IF (ABS(F) + ANORM .EQ. ANORM) GO TO 565.

There are three such tests in the subroutine.

We made this change with some reluctance because the EISPACK subroutines have been carefully and extensively tested and are distributed and supported by the NATS Project. However, the change is fairly minor, and we feel that the machine independence thereby obtained is worthwhile.

The input to SVD is an m-by-n matrix A, the dimension information, and two logical variables which indicate whether or not the matrices U and V are to be computed. The output is the n singular values (usually, but not necessarily, in decreasing order), an m-by-n matrix U if desired, and an n-by-n matrix V if desired. To save storage on large problems, either U or V can overwrite A. The Golub and Reinsch article indicates how the procedure can be easily modified to produce the full m-by-m U.

The following sample program involves the example we have used throughout the chapter. The singular values are 35.127223, 2.465397, and 0.000000. The matrices U and V are given near the beginning of the chapter. Only the first three columns of U are produced, and on some computers the signs of columns of both U and V may be changed.

```
C  SAMPLE PROGRAM FOR SVD
C
      REAL  A(5,3),U(5,3),V(5,3),SIGMA(3),WORK(5)
      INTEGER  I, IERR, J, M, N, NM
      NM = 5
      M = 5
      N = 3
      DO 1 I = 1, M
      DO 1 J = 1, N
        A(I,J) = I + (J-1)*M
    1 CONTINUE
      CALL SVD(NM,M,N,A,SIGMA,.TRUE.,U,.TRUE.,V,IERR,WORK)
      IF (IERR.NE.0) WRITE(6,2) IERR
    2 FORMAT(15H TROUBLE. IERR=, I4)
      DO 3 J = 1, N
        WRITE(6,6) SIGMA(J)
    3 CONTINUE
      WRITE(6,7)
      DO 4 I = 1, M
        WRITE(6,6) (U(I,J), J=1,N)
    4 CONTINUE
      WRITE(6,7)
      DO 5 I = 1, N
        WRITE(6,6) (V(I,J), J=1,N)
    5 CONTINUE
    6 FORMAT(3F10.6)
    7 FORMAT(1H )
      STOP
      END
```

```
      SUBROUTINE SVD(NM,M,N,A,W,MATU,U,MATV,V,IERR,RV1)
C
      INTEGER I,J,K,L,M,N,II,I1,KK,K1,LL,L1,MN,NM,ITS,IERR
      REAL A(NM,N),W(N),U(NM,N),V(NM,N),RV1(N)
      REAL C,F,G,H,S,X,Y,Z,SCALE,ANORM
      LOGICAL MATU,MATV
C
C     THIS SUBROUTINE IS A TRANSLATION OF THE ALGOL PROCEDURE SVD,
C     NUM. MATH. 14, 403-420(1970) BY GOLUB AND REINSCH.
C     HANDBOOK FOR AUTO. COMP., VOL II-LINEAR ALGEBRA, 134-151(1971).
C
C     THIS SUBROUTINE DETERMINES THE SINGULAR VALUE DECOMPOSITION
C          T
C     A=USV  OF A REAL M BY N RECTANGULAR MATRIX.  HOUSEHOLDER
C     BIDIAGONALIZATION AND A VARIANT OF THE QR ALGORITHM ARE USED.
C
C     ON INPUT.
C
C        NM MUST BE SET TO THE ROW DIMENSION OF TWO-DIMENSIONAL
C          ARRAY PARAMETERS AS DECLARED IN THE CALLING PROGRAM
C          DIMENSION STATEMENT.  NOTE THAT NM MUST BE AT LEAST
C          AS LARGE AS THE MAXIMUM OF M AND N.
C
C        M IS THE NUMBER OF ROWS OF A (AND U).
C
C        N IS THE NUMBER OF COLUMNS OF A (AND U) AND THE ORDER OF V.
C
C        A CONTAINS THE RECTANGULAR INPUT MATRIX TO BE DECOMPOSED.
C
C        MATU SHOULD BE SET TO .TRUE. IF THE U MATRIX IN THE
C          DECOMPOSITION IS DESIRED, AND TO .FALSE. OTHERWISE.
C
C        MATV SHOULD BE SET TO .TRUE. IF THE V MATRIX IN THE
C          DECOMPOSITION IS DESIRED, AND TO .FALSE. OTHERWISE.
C
C     ON OUTPUT.
C
C        A IS UNALTERED (UNLESS OVERWRITTEN BY U OR V).
C
C        W CONTAINS THE N (NON-NEGATIVE) SINGULAR VALUES OF A (THE
C          DIAGONAL ELEMENTS OF S).  THEY ARE UNORDERED.  IF AN
C          ERROR EXIT IS MADE, THE SINGULAR VALUES SHOULD BE CORRECT
C          FOR INDICES IERR+1,IERR+2,...,N.
C
C        U CONTAINS THE MATRIX U (ORTHOGONAL COLUMN VECTORS) OF THE
C          DECOMPOSITION IF MATU HAS BEEN SET TO .TRUE.  OTHERWISE
C          U IS USED AS A TEMPORARY ARRAY.  U MAY COINCIDE WITH A.
C          IF AN ERROR EXIT IS MADE, THE COLUMNS OF U CORRESPONDING
C          TO INDICES OF CORRECT SINGULAR VALUES SHOULD BE CORRECT.
```

```
C
C          V CONTAINS THE MATRIX V (ORTHOGONAL) OF THE DECOMPOSITION IF
C          MATV HAS BEEN SET TO .TRUE.  OTHERWISE V IS NOT REFERENCED.
C          V MAY ALSO COINCIDE WITH A IF U IS NOT NEEDED.  IF AN ERROR
C          EXIT IS MADE, THE COLUMNS OF V CORRESPONDING TO INDICES OF
C          CORRECT SINGULAR VALUES SHOULD BE CORRECT.
C
C          IERR IS SET TO
C            ZERO       FOR NORMAL RETURN,
C            K          IF THE K-TH SINGULAR VALUE HAS NOT BEEN
C                       DETERMINED AFTER 30 ITERATIONS.
C
C          RV1 IS A TEMPORARY STORAGE ARRAY.
C
C     QUESTIONS AND COMMENTS SHOULD BE DIRECTED TO B. S. GARBOW,
C     APPLIED MATHEMATICS DIVISION, ARGONNE NATIONAL LABORATORY
C
C     MODIFIED TO ELIMINATE MACHEP
C
      IERR = 0
C
      DO 100 I = 1, M
C
        DO 100 J = 1, N
        U(I,J) = A(I,J)
  100 CONTINUE
C     .......... HOUSEHOLDER REDUCTION TO BIDIAGONAL FORM ..........
      G = 0.0
      SCALE = 0.0
      ANORM = 0.0
C
      DO 300 I = 1, N
        L = I + 1
        RV1(I) = SCALE * G
        G = 0.0
        S = 0.0
        SCALE = 0.0
        IF (I .GT. M) GO TO 210
C
        DO 120 K = I, M
  120   SCALE = SCALE + ABS(U(K,I))
C
        IF (SCALE .EQ. 0.0) GO TO 210
C
        DO 130 K = I, M
          U(K,I) = U(K,I) / SCALE
          S = S + U(K,I)**2
  130   CONTINUE
```

```
C
      F = U(I,I)
      G = -SIGN(SQRT(S),F)
      H = F * G - S
      U(I,I) = F - G
      IF (I .EQ. N) GO TO 190
C
      DO 150 J = L, N
        S = 0.0
C
        DO 140 K = I, M
140     S = S + U(K,I) * U(K,J)
C
        F = S / H
C
        DO 150 K = I, M
          U(K,J) = U(K,J) + F * U(K,I)
150   CONTINUE
C
190   DO 200 K = I, M
200   U(K,I) = SCALE * U(K,I)
C
210   W(I) = SCALE * G
      G = 0.0
      S = 0.0
      SCALE = 0.0
      IF (I .GT. M .OR. I .EQ. N) GO TO 290
C
      DO 220 K = L, N
220   SCALE = SCALE + ABS(U(I,K))
C
      IF (SCALE .EQ. 0.0) GO TO 290
C
      DO 230 K = L, N
        U(I,K) = U(I,K) / SCALE
        S = S + U(I,K)**2
230   CONTINUE
C
      F = U(I,L)
      G = -SIGN(SQRT(S),F)
      H = F * G - S
      U(I,L) = F - G
C
      DO 240 K = L, N
240   RV1(K) = U(I,K) / H
C
      IF (I .EQ. M) GO TO 270
C
```

```
          DO 260 J = L, M
             S = 0.0
C
          DO 250 K = L, N
250          S = S + U(J,K) * U(I,K)
C
          DO 260 K = L, N
             U(J,K) = U(J,K) + S * RV1(K)
260       CONTINUE
C
270       DO 280 K = L, N
280       U(I,K) = SCALE * U(I,K)
C
290       ANORM = AMAX1(ANORM,ABS(W(I))+ABS(RV1(I)))
300    CONTINUE
C      .......... ACCUMULATION OF RIGHT-HAND TRANSFORMATIONS ..........
       IF (.NOT. MATV) GO TO 410
C      .......... FOR I=N STEP -1 UNTIL 1 DO -- ..........
       DO 400 II = 1, N
          I = N + 1 - II
          IF (I .EQ. N) GO TO 390
          IF (G .EQ. 0.0) GO TO 360
C
          DO 320 J = L, N
C      .......... DOUBLE DIVISION AVOIDS POSSIBLE UNDERFLOW ..........
320       V(J,I) = (U(I,J) / U(I,L)) / G
C
          DO 350 J = L, N
             S = 0.0
C
          DO 340 K = L, N
340          S = S + U(I,K) * V(K,J)
C
          DO 350 K = L, N
             V(K,J) = V(K,J) + S * V(K,I)
350       CONTINUE
C
360       DO 380 J = L, N
             V(I,J) = 0.0
             V(J,I) = 0.0
380       CONTINUE
C
390       V(I,I) = 1.0
          G = RV1(I)
          L = I
400    CONTINUE
C      .......... ACCUMULATION OF LEFT-HAND TRANSFORMATIONS ..........
410    IF (.NOT. MATU) GO TO 510
C      ..........FOR I=MIN(M,N) STEP -1 UNTIL 1 DO -- ..........
       MN = N
       IF (M .LT. N) MN = M
```

```
C
      DO 500 II = 1, MN
          I = MN + 1 - II
          L = I + 1
          G = W(I)
          IF (I .EQ. N) GO TO 430
C
          DO 420 J = L, N
  420     U(I,J) = 0.0
C
  430     IF (G .EQ. 0.0) GO TO 475
          IF (I .EQ. MN) GO TO 460
C
          DO 450 J = L, N
              S = 0.0
C
              DO 440 K = L, M
  440         S = S + U(K,I) * U(K,J)
C         ......... DOUBLE DIVISION AVOIDS POSSIBLE UNDERFLOW .........
              F = (S / U(I,I)) / G
C
              DO 450 K = I, M
                  U(K,J) = U(K,J) + F * U(K,I)
  450     CONTINUE
C
  460     DO 470 J = I, M
  470     U(J,I) = U(J,I) / G
C
          GO TO 490
C
  475     DO 480 J = I, M
  480     U(J,I) = 0.0
C
  490     U(I,I) = U(I,I) + 1.0
  500 CONTINUE
C         ......... DIAGONALIZATION OF THE BIDIAGONAL FORM .........
C         ......... FOR K=N STEP -1 UNTIL 1 DO -- .........
  510 DO 700 KK = 1, N
          K1 = N - KK
          K = K1 + 1
          ITS = 0
C         ......... TEST FOR SPLITTING.
C                   FOR L=K STEP -1 UNTIL 1 DO -- .........
  520     DO 530 LL = 1, K
              L1 = K - LL
              L = L1 + 1
              IF (ABS(RV1(L)) + ANORM .EQ. ANORM) GO TO 565
C         ......... RV1(1) IS ALWAYS ZERO, SO THERE IS NO EXIT
C                   THROUGH THE BOTTOM OF THE LOOP .........
              IF (ABS(W(L1)) + ANORM .EQ. ANORM) GO TO 540
  530     CONTINUE
```

```
C        ......... CANCELLATION OF RV1(L) IF L GREATER THAN 1 .........
  540    C = 0.0
         S = 1.0
C
         DO 560 I = L, K
           F = S * RV1(I)
           RV1(I) = C * RV1(I)
           IF (ABS(F) + ANORM .EQ. ANORM) GO TO 565
           G = W(I)
           H = SQRT(F*F+G*G)
           W(I) = H
           C = G / H
           S = -F / H
           IF (.NOT. MATU) GO TO 560
C
           DO 550 J = 1, M
             Y = U(J,L1)
             Z = U(J,I)
             U(J,L1) = Y * C + Z * S
             U(J,I) = -Y * S + Z * C
  550      CONTINUE
C
  560    CONTINUE
C        ......... TEST FOR CONVERGENCE .........
  565    Z = W(K)
         IF (L .EQ. K) GO TO 650
C        ......... SHIFT FROM BOTTOM 2 BY 2 MINOR .........
         IF (ITS .EQ. 30) GO TO 1000
         ITS = ITS + 1
         X = W(L)
         Y = W(K1)
         G = RV1(K1)
         H = RV1(K)
         F = ((Y - Z) * (Y + Z) + (G - H) * (G + H)) / (2.0 * H * Y)
         G = SQRT(F*F+1.0)
         F = ((X - Z) * (X + Z) + H * (Y / (F + SIGN(G,F)) - H)) / X
C        ......... NEXT QR TRANSFORMATION .........
         C = 1.0
         S = 1.0
C
         DO 600 I1 = L, K1
           I = I1 + 1
           G = RV1(I)
           Y = W(I)
           H = S * G
           G = C * G
           Z = SQRT(F*F+H*H)
           RV1(I1) = Z
           C = F / Z
           S = H / Z
           F = X * C + G * S
           G = -X * S + G * C
```

```
                  H = Y * S
                  Y = Y * C
                  IF (.NOT. MATV) GO TO 575
C
                  DO 570 J = 1, N
                     X = V(J,I1)
                     Z = V(J,I)
                     V(J,I1) = X * C + Z * S
                     V(J,I) = -X * S + Z * C
  570         CONTINUE
C
  575         Z = SQRT(F*F+H*H)
              W(I1) = Z
C    .......... ROTATION CAN BE ARBITRARY IF Z IS ZERO ..........
              IF (Z .EQ. 0.0) GO TO 580
              C = F / Z
              S = H / Z
  580         F = C * G + S * Y
              X = -S * G + C * Y
              IF (.NOT. MATU) GO TO 600
C
              DO 590 J = 1, M
                 Y = U(J,I1)
                 Z = U(J,I)
                 U(J,I1) = Y * C + Z * S
                 U(J,I) = -Y * S + Z * C
  590         CONTINUE
C
  600     CONTINUE
C
          RV1(L) = 0.0
          RV1(K) = F
          W(K) = X
          GO TO 520
C    .......... CONVERGENCE ..........
  650     IF (Z .GE. 0.0) GO TO 700
C    .......... W(K) IS MADE NON-NEGATIVE ..........
          W(K) = -Z
          IF (.NOT. MATV) GO TO 700
C
          DO 690 J = 1, N
  690     V(J,K) = -V(J,K)
C
  700 CONTINUE
C
      GO TO 1001
C    .......... SET ERROR -- NO CONVERGENCE TO A
C                    SINGULAR VALUE AFTER 30 ITERATIONS ..........
 1000 IERR = K
 1001 RETURN
      END
```

PROBLEMS

P9-1. Generate 11 data points by taking $t_i = (i - 1)/10$ and $y_i = \text{erf}(t_i)$, $i = 1, \ldots, 11$. Compute erf (t) using any of the techniques from Problems P1-1, P5-1, or P6-1.

(a) Fit the data in a least-squares sense by polynomials of degrees from 1 to 10. Compare the fitted polynomial with erf (t) for a number of values of t in between the data points, and see how the maximum error depends on n, the number of coefficients in the polynomial.

(b) Since erf (t) is an odd function of t, that is, erf $(t) = -\text{erf}(-t)$, it is reasonable to fit the same data by a linear combination of odd powers of t,

$$\text{erf}(t) \approx c_1 t + c_2 t^3 + \ldots + c_n t^{2n-1}.$$

Again, see how the error in between data points depends on n. Since t varies over $[0, 1]$ in this problem, it is not necessary to consider using other basis polynomials.

(c) Polynomials are not particularly good approximants for erf (t) because they are unbounded for large t, whereas erf (t) approaches 1 for large t. So, using the same data points, fit a model of the form

$$\text{erf}(t) \approx c_1 + e^{-t^2}(c_2 + c_3 z + c_4 z^2 + c_5 z^3)$$

where $z = 1/(1 + t)$. How does the error in between data points compare with the polynomial models?

(d) Fit erf (t) by the same model as in part (c) but with

$$z = \frac{1}{1 + \lambda t},$$

where λ is a nonlinear parameter which can be found using FMIN as described in Problem P9-3.

P9-2. Fit the $m = 7$ data points in Problem P4-7 by polynomials of degree 0 through 6. Graph the resulting polynomials as well as the interpolatory spline fit.

P9-3. *Separable least squares*

Suppose m data points (t_i, y_i), $i = 1, \ldots, m$, are to be fitted in the least-squares sense by the following function of t:

$$y(t) = c_1 + c_2 t + c_3 t^2 + c_4 e^{\lambda t}.$$

The function involves five parameters: c_1, c_2, c_3, c_4, and λ. The four c's are involved linearly, but λ is involved in a nonlinear way.

Let $A(\lambda)$ be the m-by-4 matrix with elements

$$a_{i,j} = t_i^{j-1}, \quad j = 1, 2, 3,$$
$$a_{i,4} = e^{\lambda t_i}.$$

Let y be the given m vector of data y_i, and let c be the unknown 4 vector of coefficients c_j. We then have the following optimization problem:

$$\min_{\lambda} \min_{c} \| A(\lambda)c - y \|^2.$$

The *inner* minimum involving the four linear parameters can be found for any λ using SVD. The *outer* minimum involving the single nonlinear parameter λ can be found using FMIN. Note that SVD will be called by the function subprogram called by FMIN.

Here are two sets of data on which to carry out this technique. The first set should present no difficulty, but the second set leads to a degeneracy which can be detected by monitoring the value of the inner minimum and the singular values of $A(\lambda)$. Find the c's and the λ which give a best fit. Are the c's unique? Assume the data are accurate to only two decimal places, so that the y_i may have errors as large as 0.005.

t	y, First Set	y, Second Set
0.0	20.00	20.00
0.25	51.58	24.13
0.50	68.73	26.50
0.75	75.46	27.13
1.00	74.36	26.00
1.25	67.09	23.13
1.50	54.73	18.50
1.75	37.98	12.13
2.00	17.28	4.00

P9-4. (a) Fit the census data from Problem P4-5 by various degree polynomials. Use the fits to predict the 1980 population. How is the predicted population affected by your choice of basis polynomials? By your choice of tolerance for negligible singular values? By the precision of arithmetic if you have a choice?

(b) Fit the census data by

$$y(t) \approx c_1 + c_2(t - 1900) + c_3 e^{\lambda(t-1900)}$$

with variable λ as described in Problem P9-3. Predict the 1980 population.

(c) Try to fit the census data by a quadratic

$$y(t) \approx c_1 + c_2 t + c_3 t^2$$

using the normal equations approach. What is the condition number of the resulting matrix? What is the predicted 1980 population?

P9-5.   *Singular value analysis of cryptograms*

This problem is based on a paper of Moler and Morrison (1977). The objective is to separate vowels from consonants in coded messages.

Have someone prepare a cryptogram by taking a text of several hundred characters, omitting the blanks and punctuation marks, and carrying out a simple one-to-one substitution for the letters. For example, the original text might be

NOW IS THE TIME FOR MEN TO AID THEIR COUNTRY.

Using the "2001" code where each letter is replaced by its predecessor in the alphabet (recall that the computer in Stanley Kubrick's "2001" is named HAL), the coded message is

MNVHRSGDSHLDENQLDMSNZHCSGDHQBNTMSQX.

Write a program which processes such a text and forms a 26-by-26 matrix whose elements are thought of as indexed by letters rather than integers. The (x, y)th element is the number of times the pair "XY" occurs in the coded text. In the example, $a_{m.n} = 1$, $a_{n,v} = 1, \ldots, a_{s,g} = 2$, etc. The last character of the text is thought of as being followed by the first, so $a_{x,m} = 1$. This matrix is called the *digraph frequency matrix*.

Obtain the SVD of the digraph frequency matrix.

Print out the components of the columns of U and V which correspond to the largest singular value. Ordinarily these will be $u_{x,1}$ and $v_{x,1}$, $x = a$, $b, \ldots, z$. The components of these vectors should all have the same sign and should roughly be proportional to the frequencies of the individual letters. So in an uncoded message with typical distribution of letters, the eth component will be the largest, the tth the next largest, and so on. Any missing letters will have components near zero.

Now consider the components of the columns of U and V which correspond to the *second* singular value. If the singular values are ordered, these will be $u_{x,2}$ and $v_{x,2}$. Associate with each letter, x, the point in the two-dimensional plane whose cartesian coordinates are the xth components of these vectors, that is, the point $(u_{x,2}, v_{x,2})$. Since the components are in the interval $[-1, 1]$, this will produce 26 points within the square of side 2 centered at the origin.

Most of the letters should be associated with points in the second and fourth quadrants of the plane. In other words, there will be only a few letters whose components in the second right and left singular vectors have the same sign. Furthermore, most of the consonants should be in one quadrant and most of the vowels in the other. This is related to the fact that vowel-consonant and consonant-vowels pairs are more frequent than vowel-vowel and consonant-consonant pairs.

There will be some exceptions. For example, the letter H is usually preceded by a consonant, namely T, and followed by a vowel. The letter N is often followed by another consonant. In fact, such exceptions may help in identifying these letters in the cryptogram.

It is also interesting to use texts in languages other than English. For example, German does not work so well because it contains relatively more vowel-vowel and consonant-consonant pairs.

P9-6. *Pseudoinverse*
Write a subroutine

SUBROUTINE PSEUDO (NM, M, N, A, TOL, APLUS, WORK)

that uses the SVD to compute the effective pseudoinverse of the *m*-by-*n* matrix A. Singular values less than TOL should be regarded as negligible. Test the subroutine on the example given in this chapter and some other matrices of your choice. Find an example for which changing TOL produces several different results.

The part of your subroutine that actually forms $A^\dagger$ might look something like this:

```
           DO 20 I = 1, N
           DO 20 J = 1, M
             T = 0.0
             DO 10 K = 1, N
               IF (SIGMA(K) .GT. TOL)
      *          T = T + V(I,K)*U(J,K)/SIGMA(K)
      10     CONTINUE
             APLUS(I,J) = T
      20   CONTINUE.
```

Try to organize your program so that as little extra storage outside of A and APLUS is required as possible.

P9-7. *Pseudoinverse*
Prove that the effective pseudoinverse satisfies the four conditions

$$XAX = X;$$

AX is symmetric;

XA is symmetric;

$$\|AXA - A\| \leq \tau.$$

10

RANDOM NUMBER GENERATION AND MONTE CARLO METHODS

There are many problems in which a random element is naturally present. For example, if you want to study the effect of the number of butchers on the quality of service in a butcher shop, you need to model the unpredictable element in the times at which customers enter the shop. Random noise is another example. Many types of failures in a communication system are best modeled as random effects.

There are other problems which do not by themselves have a random element but where it is useful to take random samples for one purpose or another. For example, in the decennial census of the United States, it is both too expensive and unnecessary to give every person the most complete set of questions. It is customary to reserve the complete set for such a sample as 1 person in 10, or 1 person in 40, where the sample is selected at random to avoid any systematic bias. (Every tenth house might be on the corner, and thereby select richer people.)

Another problem where random sampling turns out to be surprisingly useful is the rough evaluation of volumes of complicated regions in Euclidean space of more than four or five dimensions. This includes the approximate evaluation of integrals, of course. Let the region be R; it is usually defined by a number of inequality conditions. Suppose that R is a subset of the unit cube K in n dimensions. The *Monte Carlo* idea is to select a large number N of points in K at random, with equal probability of being in any part of K. Then one counts the number, say M, that fall inside R, i.e., that satisfy the inequality conditions defining R. Then M/N is an estimate of the volume of R. Students familiar with the binomial distribution will realize that the variance of the estimate M/N is rather high, so the accuracy of the estimate is low. Still, a sample of 10,000 points will yield an accuracy of something like 1% if the volume is not too close to 0 or 1. This accuracy is often sufficient, and it may be quite difficult to do better by other methods.

For any of these applications on a computer it is necessary to produce a large sequence of numbers that behave like random samples from a given distribution. There are three classes of methods for producing such random numbers:

1. One can use a random physical process, like the time when a hot cathode emits an electron, interpreted as the phase in some clock cycle. Such effects are easily produced, but timing them is very difficult without introducing a bias into the numbers that will have to be counteracted by other methods. In any case, the physical process, being random, cannot be reproduced, and so it is difficult to debug a computer program using them.

2. One can compute the random numbers off line and store them in a disk or tape file, to be consumed by a computer program as needed. This is possible, and this idea was much used in precomputer days when several books of tables of random numbers were published. It leaves open the question of how the file of numbers is produced. If you as a programmer did not produce your own file, then probably an existing file would be overused. One possible solution is to start using the file at a random location, but this brings us back to the problem of how to generate a random location in the file. Elaborate methods for entering a table of random numbers have been devised by statisticians in the past.

3. The third and commonest method is to use an algorithm on line with your computer program, to generate "random" numbers. These are available when they are needed and are perfectly reproducible for program checkout. Since such numbers are generated by an algorithm, they are not random at all and should be called *pseudorandom*. Nevertheless, this approach seems to be the most useful and will be the only approach discussed in these notes.

10.1. GENERATION OF UNIFORMLY DISTRIBUTED NUMBERS

Methods for generating random numbers from one or another distribution are generally based on first getting a random real number from a uniform distribution on the interval [0, 1). Such a number is later converted to another distribution. Because integer arithmetic is better understood, one starts by getting an integer that is uniformly distributed over the range of nonnegative integers; for example, on System/360, that means between 0 and $2^{31} - 1$. Such an integer can be transformed into a floating-point number on [0, 1). (However, it is considerably faster to avoid the conversion and multiplication, and for many purposes the random integer is just as useful as the random real number.)

The almost universally used method of generating pseudorandom integers is to select a function f that maps the integers into themselves. Select x_0 somehow, and generate the next integer by the function $x_{k+1} = f(x_k)$. Initially functions f were selected by being as complicated and confusing and little understood as possible—for example, $f(x)$ was the integer whose binary representation was the middle 31 digits of the 62-digit square of x. But the lack of a theory about f proves disastrous. In the midsquare method, $f(x)$ can occasionally turn out to be 0 at unpredictable times, and then all following x_k are 0. So quite early people changed to the use of functions f whose properties were completely understood. Any sequence of integers out of the set $(0, 2^{31} - 1)$ has to repeat after at most $2^{31} \approx 10^9$ elements. With the aid of number theory, it is possible to choose f so that the period is known a priori to be as long, or nearly as long, as possible. This avoids premature termination or cycling of the sequence. The further use of number theory can more or less predict the character of the sequence, to give the user some degree of confidence that it will serve well enough to simulate a random sequence of numbers.

The commonest function f now used takes the form

$$f(x) = ax + c(\text{modulo } m),$$

where m is usually 2^t for t-digit binary integers. Here x_0, a, and c are themselves integers in the same range. Some choices of a and c are good, and some are not. (*Example:* $a = c = 1$ would obviously be awful.) Knuth (1969) summarizes the number theory needed to pick a and c; see pp. 78 and 155:

1. x_0 can be arbitrary. Perhaps $x_0 = 1$ for program checkout, and x_0 is a digitized version of the clock time for production running, to yield different sequences for different runs.

2. Pick a to have three properties (for binary machines):
   a. $a \bmod 8 = 5$;
   b. $m/100 < a < m - \sqrt{m}$;
   c. The binary digits of a have no obvious pattern.

3. Pick c as an odd integer with

$$\frac{c}{m} \approx \frac{1}{2} - \frac{1}{6}\sqrt{3} \approx 0.21132.$$

One must remember that the least significant binary digits of the x_k will not be very random. Hence, for example, if you want to choose at random

1 of 16 possible branches by use of x_k, use the *most* significant bits of x_k, not the least significant ones. Finally, for greatest safety, it is wise to *pretest* random numbers by trying them out for some problem similar to the real application for which one knows the answers.

We have programmed a simple uniform random generator following the suggestions of Knuth. The subroutine is presented at the end of this section in ANSI standard Fortran. Since the subroutine is designed to be relatively machine independent, we call it URAND, which stands for "universal <u>ran</u>dom number generator" and for "<u>uniform random</u> number generator."

URAND produces a sequence of integers by setting

$$Y_{n+1} = aY_n + c \text{ (modulo } m), \qquad n \geq 1,$$

on the nth call of URAND. These are converted into floating-point numbers in the interval $[0, 1)$ and returned as the value of URAND. The resulting value of Y_{n+1} is returned through the parameter IY and should be used for the actual parameter in the subsequent call. On the first call of URAND, IY should be initialized to an arbitrary integer value.

The values of m, a, and c are computed automatically upon the initial entry. The main assumption here is that the machine uses binary integer number representation and multiplication is performed modulo m, where m is a power of 2. This assumption simplifies the computation. URAND discovers the value of $m/2$ by testing successive powers of 2 until a multiplication by 2 produces no increase in magnitude. It is also assumed that integer addition is either modulo m or that at least $\log_2(m)$ significant bits are returned. The values of a and c are computed following the advice of Knuth, outlined above. In the source code, a is called IA, and c is called IC. The random bit pattern of a is achieved by calling DATAN(1.D0), which returns the double-precision value of $\pi/4$, which, on a binary machine, is the shifted bit pattern of π. The division by 8.D0 and multiplication by $m/2$ is hopefully accomplished without unduly altering this pattern. The double-precision value is finally converted to an integer, multiplied by 8, and incremented by 5 to ensure a mod $8 = 5$. The resulting value of a is roughly $(m/8)\pi \approx m/2$. This satisfies the inequality constraints. The value of c is computed directly from the definition 3. Some Fortran compilers do not convert constants like 8.D0 to exact floating-point representations, but this problem will probably be of little consequence.

The sequence $[Y_n]$ is guaranteed to have maximum period length m by Theorem A given in Knuth (1969), p. 15. However, the least significant binary digits of the Y_n will not be very random. When the Y_n are converted to floating-point numbers, the least significant digits are usually not important. To compute a random integer between 0 and K-1, one should use IFIX(K*URAND(IY)).

With their Fortran system on System/360, IBM furnished SSP, a scientific subroutine package of programs that are automatically callable from the Fortran compilers. One of these is a random number generator called RANDU. *We urge you to use URAND and not to use RANDU,* for reasons that we shall now explain.

RANDU uses $a = 65539$ and $c = 0$ in the generation of Y_n. The period of the sequence is a very satisfactory 2^{29}, ranging through one quarter of all integers. (The period of the sequence produced by URAND on the IBM 360 is 2^{31}, as long as possible.) The difficulty comes from the fact that $65539 = 2^{16} + 3$. Perhaps this number was chosen for fast multiplication, but it leads to disastrous properties of the sequence. The following relations are all taken modulo 2^{31}:

$$x_{k+2} = (2^{16} + 3)x_{k+1} = (2^{16} + 3)^2 x_k$$
$$= (2^{32} + 6 \cdot 2^{16} + 9)x_k = [6 \cdot (2^{16} + 3) - 9]x_k$$
$$= 6x_{k+1} - 9x_k.$$

Hence,

$$x_{k+2} = 6x_{k+1} - 9x_k, \qquad \text{for all } k.$$

As a result, there is an extremely high correlation among three successive random integers of the sequence generated by RANDU. For example, each time that x_k is near 0, x_{k+2} is close to six times x_{k+1}, modulo 2^{31}.

We have tested 10,000 to 100,000 triples of numbers (x_k, x_{k+1}, x_{k+2}) generated by both RANDU and URAND. The triples are put into 1000 boxes, according to which of the tenths of the interval $[0, 2^{31} - 1)$ each of the three variables belongs to. (The interval $[0, 1)$ is divided into ten equal intervals for testing URAND.) If the random integer generator were truly random, each triple should have an equal chance of falling into any of the 1000 boxes. The distribution of the triples over the boxes can be given a chi-square test, as described by Knuth (1969). The results showed that the URAND was perfectly satsifactory but that RANDU was so bad as to be totally discredited as a random number generator. But we believe RANDU to be so bad that many simulations using it probably have demonstrably bad behavior, especially where triples of numbers are used. Since RANDU is almost universally used by System/360 users, we seriously question the results of a large number of simulations.

Marsaglia (1968) has proved that to a certain degree all random number generators using recurrences will suffer correlations among successive numbers, but for well-designed generators, like URAND, the effect is far smaller than for many others.

Many other tests of random number generators have been devised, but there is ordinarily none better than using them in a model problem in your own field of application, with answers that are known to you.

There is a growing literature on methods for determining sequences of numbers x_k on [0, 1) that do not have a random look but which are well suited for a given application. For example, there are sequences of x_k which are equidistributed over the interval [0, 1) with much more regularity than truly random numbers. These may be useful for Monte Carlo integration because of a reduced variance of the estimates.

If the speed of generating uniformly distributed random numbers is vitally important in your program, you should examine specialized programs for your machine published in the recent literature. For example, there is a Fortran program by Marsaglia and Bray (1968) and a machine-language program for System/360 by Seraphin (1969). The time for the machine-language subroutine seems to be 11.9 microseconds on the 360/67, plus 4.2 microseconds to call the subroutine. URAND takes around 90 microseconds in Fortran (H), (OPT = 2) on the IBM 360/67.

10.2. SUBROUTINE URAND

The following program illustrates a trivial use of URAND. It prints ten lines, each line containing a random "heads" or "tails."

```
C  SAMPLE PROGRAM FOR URAND
C
      IY = 0
      DO 10 I=1,10
         IF (URAND(IY) .LT. 0.5) GO TO 5
         WRITE (6,1)
         GO TO 10
    5    WRITE (6,2)
   10 CONTINUE
    1 FORMAT(6H HEADS)
    2 FORMAT(6H TAILS)
      STOP
      END
```

A different starting value of the variable IY will produce a different sequence of random numbers. Moreover, even with the same starting value for IY, different sequences will be produced on computers with different word sizes.

```
      REAL FUNCTION URAND(IY)
      INTEGER  IY
C
C     URAND IS A UNIFORM RANDOM NUMBER GENERATOR BASED  ON   THEORY  AND
C SUGGESTIONS  GIVEN  IN  D.E. KNUTH (1969),  VOL  2.   THE INTEGER  IY
C SHOULD BE INITIALIZED TO AN ARBITRARY INTEGER PRIOR TO THE FIRST CALL
C TO URAND.  THE CALLING PROGRAM SHOULD  NOT  ALTER  THE  VALUE  OF  IY
C BETWEEN  SUBSEQUENT CALLS TO URAND.  VALUES OF URAND WILL BE RETURNED
C IN THE INTERVAL (0,1).
C
      INTEGER   IA,IC,ITWO,M2,M,MIC
      DOUBLE PRECISION   HALFM
      REAL  S
      DOUBLE PRECISION   DATAN,DSQRT
      DATA M2/0/,ITWO/2/
      IF (M2 .NE. 0) GO TO 20
C
C IF FIRST ENTRY, COMPUTE MACHINE INTEGER WORD LENGTH
C
      M = 1
   10 M2 = M
      M = ITWO*M2
      IF (M .GT. M2) GO TO 10
      HALFM = M2
C
C COMPUTE MULTIPLIER AND INCREMENT FOR LINEAR CONGRUENTIAL METHOD
C
      IA = 8*IDINT(HALFM*DATAN(1.D0)/8.D0) + 5
      IC = 2*IDINT(HALFM*(0.5D0-DSQRT(3.D0)/6.D0)) + 1
      MIC = (M2 - IC) + M2
C
C S IS THE SCALE FACTOR FOR CONVERTING TO FLOATING POINT
C
      S = 0.5/HALFM
C
C COMPUTE NEXT RANDOM NUMBER
C
   20 IY = IY*IA
C
C THE FOLLOWING STATEMENT IS FOR COMPUTERS WHICH DO NOT ALLOW
C INTEGER OVERFLOW ON ADDITION
C
      IF (IY .GT. MIC) IY = (IY - M2) - M2
C
      IY = IY + IC
C
C THE FOLLOWING STATEMENT IS FOR COMPUTERS WHERE THE
C WORD LENGTH FOR ADDITION IS GREATER THAN FOR MULTIPLICATION
C
      IF (IY/2 .GT. M2) IY = (IY - M2) - M2
C
C THE FOLLOWING STATEMENT IS FOR COMPUTERS WHERE INTEGER
C OVERFLOW AFFECTS THE SIGN BIT
C
      IF (IY .LT. 0) IY = (IY + M2) + M2
      URAND = FLOAT(IY)*S
      RETURN
      END
```

10.3. SAMPLING FROM OTHER DISTRIBUTIONS

Often the generation of a random number in $[0, 1)$ is merely a means to making some random decision. As a very simple example, suppose in a simulation program it is desired to take some branch one tenth of the time, at random. An easy way to do this is to select a random number y on $[0, 1)$ and take the branch if $y < 0.1$. A faster way is to select a random integer x on $[0, 2^{31} - 1]$ and take the branch if $x < 2^{31}/10$. In the same way any finite distribution with various weights can be easily produced from uniformly distributed random integers.

Much more difficult is the computation of a random number from a non-uniform continuous distribution. For a general cumulative distribution function $F(x)$ for which special methods have not been developed, one method is to take a uniformly distributed random number on $[0, 1)$ and then form $y = F^{-1}(x)$ from the inverse function. How to form the mapping F^{-1} rapidly enough and accurately enough is the big problem.

Some common and important distributions have been much studied already. Perhaps the best known is the *normal distribution $N(0, 1)$* with mean zero and variance 1. The inverse mapping F^{-1} is relatively hard and slow to deal with. One common idea is to form 12 uniform random numbers y_i on $[0, 1)$, form $x_i = 2y_i - 1$ to get 12 uniform deviates on $[-1, 1)$, and then add them all together. The result is a fairly good approximation to a sample from $N(0, 1)$. The method is relatively slow.

A much more ingenious method is due to Box and Muller and uses the following algorithms, as later organized by James Bell of Stanford:

1. Form two uniform deviates U_1, U_2 on $[0, 1)$ (e.g., with URAND).
2. Form $V_1 = 2U_1 - 1$ and $V_2 = 2U_2 - 1$ to get two uniform deviates on $[-1, 1)$.
3. Form $S := V_1^2 + V_2^2$. If $S > 1$, discard V_1, V_2 and go to step 1. (We lose 22% of our efficiency here.) If $S \leq 1$, then we have a random point (V_1, V_2) in the unit circle.
4. Form

$$X_1 = V_1 \sqrt{\frac{-2 \ln S}{S}}, \qquad X_2 = V_2 \sqrt{\frac{-2 \ln S}{S}},$$

where ln is a natural logarithm (this takes most of the time). As proved by Knuth (1969), p. 104, X_1 and X_2 are chosen independently from the normal distribution $N(0, 1)$. A slight variant of this method has been published as an Algol 60 program by Pike (1965).

For considerably faster methods, consult Ahrens and Dieter (1972). Also, Forsythe (1972) has described a related but different method in which a random normal variable is obtained at an average cost of generating 4.036 calls on URAND, plus a few comparisons.

PROBLEMS

P10-1. If x is a random variable drawn from the normal distribution $N(0, 1)$ and z is some constant, then the probability that x is less than z, which is denoted by $P[x < z]$, can be shown to be

$$P[x < z] = \frac{1}{\sqrt{2\pi}} \int_{-\infty}^{z} e^{-t^2/2}\, dt$$

$$= \frac{1}{2}\left(1 + \operatorname{erf}\frac{z}{\sqrt{2}}\right).$$

Let M be a large interger, say 10,000. Generate M random variables x_i from $N(0, 1)$ using URAND and the Box-Muller scheme described in Section 10.3. Save them in an array. Then let z take on the values from -3.0 to 3.0 in steps of 0.2. For each z, print out and compare two quantities: (number of $x_i < z$)/M and $\frac{1}{2}(1 + \operatorname{erf}(z/\sqrt{2}))$. Obtain the values of erf using any of the techniques of Problems P1-1, P5-1, or P6-1. Try the experiment several times with different starting values for URAND. If enough computer time is available, try larger values of M.

P10-2. Use the two random number generators URAND and RANDU and carry out a chi-square test on the distribution of triples into the 1000 boxes mentioned in Section 10.1. To keep the chi-square test meaningful, be sure that each triple uses three fresh random numbers; do not try to share numbers among different triples.

P10-3. Use the Monte Carlo method to find the area of a circle of radius 1.

P10-4. Use the Monte Carlo method to find the volume in Euclidean four-dimensional space of the region satisfying all the inequalities

$$0 \le x_1 \le 1,$$
$$0 \le x_2 \le 1,$$
$$0 \le x_3 \le 1,$$
$$0 \le x_4 \le 1,$$
$$x_1 + x_2^2 + x_3^3 + x_4^4 \le 0.7,$$
$$x_1 + \sin x_4 \le 0.8,$$
$$x_1 x_3 \le 0.5.$$

P10-5. Knuth (1969) discusses a number of tests that can be used to test the effectiveness of a random number generator. Using as many of these tests as possible, try to determine the quality of the random numbers generated by URAND on your machine.

P10-6. This problem concerns random walks on lattices in one, two, and three dimensions. A random walk is a sequence of unit steps where each step is taken in the direction of one of the coordinate axes, and each possible direction has equal probability of being chosen. In two dimensions, for example, a single step starting at the point with integer coordinates (x, y) would be equally likely to move to any one of the four neighbors $(x + 1, y)$, $(x - 1, y)$, $(x, y + 1)$, $(x, y - 1)$. In one dimension there are two possible neighbors, while in three dimensions there are six possible neighbors.

URAND can be used to experimentally determine approximate answers to the following questions about random walks which start at the origin.

(a) In a one-dimensional random walk, what is the expected number of visits to some fixed nonzero point before the first return to the origin? Does the number depend on the location of the point?

(b) In a two-dimensional random walk, is there any finite bound on the expected number of steps before the first return to the origin?

(c) In a three-dimensional random walk, what is the probability of eventual return to the origin?

Some of these questions require fairly large amounts of computer time to obtain even two significant figures in the answer. All of the questions have analytic solutions. See Feller (1950) or Spitzer (1964).

Your program will involve two large integers, M = the number of random walks to be taken and N = the maximum number of steps in a single walk. When a walk reaches N steps, it is assumed that it will not return to the origin. Check out your program using fairly small values of M and N; then make some runs with larger values. It would be desirable to investigate the effect of M and N on the results, but this would require a great deal of computer time.

REFERENCES

ABRAMOWITZ, M., and I. STEGUN (editors) (1964), *Handbook of mathematical functions, graphs, and mathematical tables*. Washington, D.C.: Government Printing Office. National Bureau of Standards Applied Mathematics Series, vol. 55 also Dover, New York (1965).

AHLBERG, H. J., E. N. NILSON and J. L. WALSH (1967), *The theory of splines and their application*, New York: Academic Press.

AHRENS, J. H., and U. DIETER (1972), "Computer methods for sampling from the exponential and normal distributions," *Comm. ACM*, vol. 15, 873–882.

ANDREWS, H. C., and C. L. PATTERSON (1975), "Outer product expansions and their uses in digital image processing," *American Mathematical Monthly*, vol. 82, 1–12.

ANONYMOUS (1966), "U.S.A. standard Fortran," *USASX3.9-1966*. United States of America Standards Institute.

BRENT, R. P. (1973), *Algorithms for minimization without derivatives*. Englewood Cliffs, N.J.: Prentice-Hall.

BROWN, K. M., and S. D. CONTE (1967), "The solution of simultaneous nonlinear equations," *Proc. 22nd National Conference of ACM*, 111–114.

BULIRSCH, R., and J. STOER (1966), "Numerical treatment of ordinary differential equations by extrapolation methods," *Numer. Math.*, vol. 8, 1–13.

DANTZIG, G. B. (1963), *Linear programming and extensions*. Princeton, N.J.: Princeton University Press.

DAVIS, H. T. (1962), *Introduction to nonlinear differential and integral equations*. New York: Dover.

DEKKER, T. J. (1969), "Finding a zero by means of successive linear interpolation," in B. DEJON and P. HENRICI (editors), *Constructive aspects of the fundamental theorem of algebra*. New York: Wiley-Interscience.

DEKKER, T. J. (1971), "A floating-point technique for extending the available precision," *Numer. Math.*, vol. 18, 224–242.

ENRIGHT, W. H., and T. E. HULL(1976), "Test results on initial value methods for non-stiff ordindary differential equations," *SIAM J. Numer. Anal.*, vol. 13, 944–961.

FADDEEV, D. K., and V. N. FADDEEVA (1963), *Computational methods of linear algebra.* San Francisco: W. H. FREEMAN. (Translated from Russian by R. D. WILLIAMS.)

FADDEEVA, V. N. (1959), *Computational methods of linear algebra.* New York: Dover. (Translated by C. D. BENSTER from a Russian book of 1950.)

FEHLBERG, E. (1970), "Klassiche Runge-Kutta-formeln vierter und niedregerer ordnung mit schrittweitenkontrolle und ihre anwendung auf warmeleitungsprobleme," *Computing*, vol. 6, 61–71.

FELLER, WILLIAM (1950), *An introduction to probability theory and its applications,* New York: Wiley.

FLETCHER, R., and M. J. D. POWELL (1963), "A rapidly convergent descent method for minimization," *Computer J.*, vol. 6, 163–168.

FORSYTHE, G. E. (1969), "What is a satisfactory quadratic equation solver," in B. DEJON and P. HENRICI (editors), *Constructive aspects of the fundamental theorem of algebra.* New York: Wiley-Interscience, pp. 53–61.

FORSYTHE, G. E. (1972), "Von Neumann's comparison method for random sampling from the normal and other distributions," *Tech. Report STAN-CS-254-72.* Stanford, Calif.: Computer Science Department, Stanford University.

FORSYTHE, G. E., and C. B. MOLER (1967), *Computer solution of linear algebraic systems.* Englewood Cliffs, N.J.: Prentice-Hall.

FORSYTHE, G. E., and W. R. WASOW (1960), *Finite difference methods for partial differential equations.* New York: Wiley.

GARDNER, M. (1961), *Mathematical puzzles & diversions.* New York: Simon and Schuster.

GEAR, C. W. (1971), *Numerical initial value problems in ordinary diflerential equations.* Englewood Cliffs, N.J.: Prentice-Hall.

GILL, P. E., and W. MURRAY (1974), *Numerical methods for constrained optimization.* New York: Academic Press.

GILL, P. E., W. MURRAY, and R. A. PITFIELD (1972), "The implementation of two quasi-Newton algorithms for unconstrained optimization," *DNAC 11.* National Physical Laboratory.

GILPIN, M. E. (1972), "Enriched preditor-prey systems: theoretical stability," *Science*, vol 177, 902–904.

GOLUB, G. H., and C. REINSCH (1971), "Singular value decomposition and least squares solutions," in J. H. WILKINSON and C. REINSCH (editors), *Handbook for automatic computation*, vol. II: "Linear algebra." Heidelberg: Springer.

GRAGG, W. B., and G. W. STEWART (1976), "A stable variant of the secant method for solving nonlinear equations," *SIAM J. Numer. Anal.*, vol. 13, 889–903.

HAMMING, R. W. (1962), *Numerical methods for scientists and engineers.* New York: McGraw-Hill.

HENRICI, P. (1962), *Discrete variable methods in ordinary diflerential equations.* New York: Wiley.

HILBERT, D. (1894), "Ein betrag zur theorie des Legendre'schen polynoms," *Acta Math.*, vol. 18, 155–160.

HULL, T. E., W. H. ENRIGHT, B. M. FELLEN, and A. E. SEDGWICK (1972), "Comparing numerical methods for ordinary differential equations," *SIAM J. Numer. Anal.*, vol. 9.

JENKINS, M. A., and J. F. TRAUB (1972), "Zeros of a complex polynomial," Algorithm 419, *Comm. ACM*, vol. 15, 97–99.

JOYCE, D. C. (1971), "Survey of extrapolation processes in numerical analysis," *SIAM Review*, vol. 13, 435–490.

KELLER, H. B. (1968), *Numerical methods for two-point boundary value problems.* Waltham, Mass.: Ginn/Blaisdell.

KNUTH, D. E. (1969), "Seminumerical algorithms," *The art of computer programming*, vol. 2. Reading Mass.: Addison-Wesley.

KROGH, F. T. (1973), "On testing a subroutine for the numerical integration of ordinary differential equations", *J. Assoc. Comput. Mach.*, vol. 20, 545–562.

LANCZOS, C. (1956), *Applied analysis.* Englewood Cliffs, N.J.: Prentice-Hall,

LAWSON, C. L., and R. J. HANSON (1974), *Solving least squares problems.* Englewood Cliffs, N.J.: Prentice-Hall.

LOTKA, A. J. (1956), *Elements of mathematical biology.* New York: Dover.

LYNESS, J. N. (1969a), "Notes on the adaptive Simpson quadrature routine," *J. ACM*, vol. 16, 483–495.

LYNESS, J. N. (1969b), "The effect of inadaquate convergence criteria in automatic routines," *Computer J.*, vol. 12, 279–281.

LYNESS, J. N. (1970), "SQUANK (Simpson quadrature used adaptively noise killed)," Algorithm 379, *Comm. ACM*, vol. 13, no. 1, April, 260–262.

LYNESS, J. N., and C. B. MOLER (1967), "Numerical differentiation of analytic functions," *SIAM J. Num. Anal.*, vol. 4, 202–210.

LYNESS, J. N., and G. SANDE (1971), "Evaluation of normalized Taylor coefficients: Algorithm 413, ENTCAF and ENTCRE," *Comm. ACM*, vol. 14, 669–675.

MALCOLM, M. A. (1972), "Algorithms to reveal properties of floating-point arithmetic," *Comm. ACM*, vol. 15, 949–951.

MALCOLM, M. A., and R. B. SIMPSON (1975), "Local vs. global strategies in adaptive quadrature," *Trans. on Math. Software*, vol. 1, 129–146.

MARSAGLIA, G. (1968), "Random numbers fall mainly in the planes," *Proc. Nat. Acad. Sci.*, vol. 61, 25–28.

MARSAGLIA, G., and T. A. BRAY (1968), "One-line random number generators and their use in combinations," *Comm. ACM*, vol. 11, 757–759.

MAY, R. M. (1972), "Limit cycles in preditor-prey communities," *Science*, vol. 177, 900–902.

McKEEMAN, W. M. (1962), "Adaptive numerical integration by Simpson's rule," Algorithm 145, *Comm. ACM*, vol. 5, 604.

MILNE, W. E. (1953), *Numerical solution of differential equations.* New York: Wiley. (Available from Dover, New York.)

MOLER, C. B. (1972), "Matrix computations with Fortran and paging," *Comm. ACM*, vol. 15, no. 4, April, 268–270.

MOLER, C. B., and D. R. MORRISON (1977), "Singular value decomposition and cryptography," manuscript submitted for publication, University of New Mexico.

MOLER, C. B., and L. P. SOLOMON (1970), "Use of splines and numerical integration in geometrical acoustics," *J. Acoustical Soc. Amer.*, vol. 48, no. 3, May, 739–744.

MOSES, J. (1972), "Toward a general theory of special functions," *Comm. ACM*, vol. 15, no. 7, 550–554.

MULLER, D. E. (1956), "A method for solving algebraic equations using an automatic computer," *Math. Tables and Other Aids to Computation*, vol. 10, 208–215.

MURRAY, W. (1972), *Numerical methods for unconstrained optimization.* New York: Academic Press.

VON NEUMANN, J., and H. H. GOLDSTINE (1947), "Numerical inverting of matrices of high order," *Bull. Amer. Math. Soc.*, vol. 53, 1021–1099, and *Proc. Amer. Math. Soc.*, vol. 2 (1951), 188–202.

ORCHARD-HAYS, W. (1968), *Advanced linear-programming computing techniques.* New York: McGraw-Hill.

ORTEGA, J. M., and W. C. RHEINBOLDT (1970), *Iterative solution of nonlinear equations in several variables.* New York: Academic Press.

OSBORNE, M. R. (1968), "A new method for the integration of stiff systems of ordinary differential equations," in *Proceedings IFIP congress 68.* Amsterdam: North-Holland.

OSTROWSKI, A. (1966), *Solution of equations and systems of equations,* 2nd ed. New York: Academic Press.

PIKE, M. C. (1965), "Random normal deviate," Algorithm 267, *Comm. ACM*, vol. 8, 606.

RALSTON, A. (1965), *A first course in numerical analysis.* New York: McGraw-Hill.

RALSTON, A., and H. WILF (editors) (1960), *Mathematical methods for digital computers,* vol. 1. New York: Wiley.

RALSTON, A., and H. WILF (editors) (1967), *Mathematical methods for digital computers*, vol. 2. New York: Wiley.

REINSCH, C. (1967), "Smoothing by spline functions," *Numer. Math.*, vol. 10, 177–183.

REINSCH, C. (1971), "Smoothing by spline functions II," *Numer. Math.*, vol. 16, 451–454.

RICE, J. R. (1975), "A metalgorithm for adaptive quadrature," *J. ACM*, vol. 22, 61–82.

RICHARDSON, L. F., and J. A. GAUNT (1927), "The deferred approach to the limit," *Trans. Roy. Soc. London*, vol. 226A, 300.

RYDER, B. G. (1974), "The PFORT verifier," *Software—Practice and Experience*, vol. 4, 359–377.

SEDGWICK, A. E. (1973), "An effective variable order, variable step Adams method," Ph. D. thesis, Department of Computer Science, Tech. Report No. 53, University of Toronto.

SERAPHIN, D. S. (1969), "A fast random number generator for IBM 360," *Comm. ACM*, vol. 12, 695.

SHAMPINE, L. F. and M. K. GORDON (1975), *Computer solution of ordinary differential equations*. San Francisco: W. H. FREEMAN.

SHAMPINE, L. F., H. A. WATTS, and S. M. DAVENPORT (1976), "Solving non-stiff ordinary differential equations—the state of the art," *SIAM Review*, vol. 18, 376–441.

SMITH, B. T., J. M. BOYLE, J. DONGARRA, B. S. GARBOW, Y. IKEBE, V. C. KLEMA, and C. B. MOLER (1976), *Matrix eigensystem routines—EISPACK guide*. Heidelberg: Springer.

SPITZER, F. L. (1964), *Principles of random walk*, Princeton: Van Nostrand.

STEGUN, I. A., and M. ABRAMOWITZ (1956), "Pitfalls in computation," *J. Soc. Indust. Appl. Math.*, vol. 4, 207–219.

STEWART, G. W. (1973), *Introduction to matrix computation*. New York: Academic Press.

WENDROFF, B. (1966), *Theoretical numerical analysis*. New York: Academic Press.

WENDROFF, B. (1969), *First principles of numerical analysis: An undergraduate text*. Reading, Mass.: Addison-Wesley.

WILDE, D. J. (1964), *Optimum seeking methods*. Englewood Cliffs, N.J.: Prentice-Hall.

WILKINSON, J. H. (1963), *Rounding errors in algebraic processes*. Englewood Cliffs, N.J.: Prentice-Hall.

WILKINSON, J. H. (1965), *The algebraic eigenvalue problem*. Oxford: Clarendon Press.

WILKINSON, J. H. (1967), "Two algorithms based on successive linear interpolation," *Tech. Report STAN-CS-67-60*. Stanford, Calif.: Computer Science Department, Stanford University.

WILKINSON, J. H., and C. REINSCH (1971), *Handbook for automatic computation.* Heidelberg: Springer.

INDEX